T0224483

SpringerBriefs in Statistics

JSS Research Series in Statistics

The current research of statistics in Japan has expanded in several directions in line with recent trends in academic activities in the area of statistics and statistical sciences over the globe. The core of these research activities in statistics in Japan has been the Japan Statistical Society (JSS). This society, the oldest and largest academic organization for statistics in Japan, was founded in 1931 by a handful of pioneer statisticians and economists and now has a history of about 80 years. Many distinguished scholars have been members, including the influential statistician Hirotugu Akaike, who was a past president of JSS, and the notable mathematician Kiyosi Itô, who was an earlier member of the Institute of Statistical Mathematics (ISM), which has been a closely related organization since the establishment of ISM. The society has two academic journals: the Journal of the Japan Statistical Society (English Series) and the Journal of the Japan Statistical Society (Japanese Series). The membership of JSS consists of researchers, teachers, and professional statisticians in many different fields including mathematics, statistics, engineering, medical sciences, government statistics, economics, business, psychology, education, and many other natural, biological, and social sciences. The JSS Series of Statistics aims to publish recent results of current research activities in the areas of statistics and statistical sciences in Japan that otherwise would not be available in English; they are complementary to the two JSS academic journals, both English and Japanese. Because the scope of a research paper in academic journals inevitably has become narrowly focused and condensed in recent years, this series is intended to fill the gap between academic research activities and the form of a single academic paper. The series will be of great interest to a wide audience of researchers, teachers, professional statisticians, and graduate students in many countries who are interested in statistics and statistical sciences, in statistical theory, and in various areas of statistical applications.

More information about this subseries at http://www.springer.com/series/13497

Nobuaki Hoshino · Shuhei Mano ·
Takaaki Shimura

Editors

Pioneering Works on Distribution Theory

In Honor of Masaaki Sibuya

 Springer

Editors
Nobuaki Hoshino
School of Economics
Kanazawa University
Kanazawa, Ishikawa, Japan

Shuhei Mano
The Institute of Statistical Mathematics
Tachikawa, Tokyo, Japan

Takaaki Shimura
The Institute of Statistical Mathematics
Tachikawa, Tokyo, Japan

ISSN 2191-544X   ISSN 2191-5458 (electronic)
SpringerBriefs in Statistics
ISSN 2364-0057   ISSN 2364-0065 (electronic)
JSS Research Series in Statistics
ISBN 978-981-15-9662-9   ISBN 978-981-15-9663-6 (eBook)
https://doi.org/10.1007/978-981-15-9663-6

Preface

Statistical distribution theory is a traditional field. Hence, it might be surprising that its frontier is still lively. The present work compiled refereed articles specific to this field and showcases pioneering efforts. Although issues discussed can be unfamiliar to even statisticians, simple examples provided are often effective in helping readers understand where the frontier exists.

A non-negligible part of the tradition has resulted from Professor Masaaki Sibuya's contributions since the 1960s. Among his various interests in the field, the theory of random partitioning of a positive integer carries significant weight. The Ewens sampling formula is the "normal" distribution over the partitions of a positive integer. Along with its generalization, which is called the Pitman sampling formula, these distributions constitute a basis for understanding the combinatorial phenomena of random partitioning.

The present volume commences with Professor Sibuya's latest research on a statistical test to discriminate the Ewens sampling formula from the Pitman sampling formula (Chapter 1). He notes the fact that the tails of the lattice structure of partitions are more likely to be observed under the generalized domain of the Pitman sampling formula compared to the original Ewens sampling formula. These tails are defined in terms of a partial order, and fundamental thoughts on the treatment of partially ordered sets are presented. In fact, few studies on the use of partially ordered sets in statistics exist. We are pleased to see Professor Sibuya's pioneering work again in this honorary volume dedicated to him.

The next two chapters also discuss the Ewens sampling formula. The number of parts in the partition of a positive integer is called a length. Professor Yamato has been working on the discrete asymptotics of lengths as well as its approximation under the Ewens sampling formula, as presented in Chapter 2. He provides new proof of the convergence of length to the shifted Poisson distribution. Professor Tsukuda employed different regimes on the asymptotics of the length that the central limit theorem dictates. He considers two types of standardization of length in Chapter 3. The derived results include the error bounds of normal approximation to these types and a decay rate depending on the growth rate of the parameter of the Ewens sampling formula.

The partition of a positive integer induces a different story. Suppose that a partition of a positive integer is given. Then, the ordering of its parts generates a sequence of positive integers. In this sequence, if the adjacent integers are the same, then this pair is called level. The number of levels in the sequence obviously depends on the method of ordering, but permutations exhaust all cases. A point of interest raised by Professor Fu in Chapter 4 is the number of permutations with a specific number of levels for a given partition. This issue dates back to 1755 and has been unsolved, but he has derived an explicit formula via finite Markov chain imbedding, separating from the ordinary combinatorial approach.

The partition also appears as the index of orthogonal polynomials. The orthogonal property of the system of orthogonal polynomials depends on a weight function, which is regarded as a statistical distribution, such as the normal, gamma, or beta distributions. These distributions can be of symmetric matrix variates, e.g., the Wishart distribution, and Professor Chikuse is interested in orthogonal polynomials associated with these cases. Despite the complex structure of the orthogonal polynomials with a symmetric matrix argument, she derived recurrence relations, which can be seen in Chapter 5.

Finally, Chapter 6 considers the parameter estimations of the regular exponential family. Professors Yanagimoto and Ohnishi rigorously proved that the posterior mean of the canonical parameter vector under Jeffreys' prior is asymptotically equivalent to the maximum likelihood estimator, which characterizes Jeffreys' prior. This result itself is a mathematical elaboration rather than a pioneering work. However, they were motivated to understand a predictor that is expressed as a function of the posterior mean of the canonical parameter in reference to the asymptotic behavior of the maximum likelihood estimator. They present a step toward their vision.

These contributive articles are based on presentations at the Pioneering Workshop on Extreme Value and Distribution Theories in Honor of Professor Masaaki Sibuya held at the Institute of Statistical Mathematics (ISM), Tokyo, Japan from March 21–23, 2019. This workshop was sponsored by Kanazawa University and the ISM and co-sponsored by the Advanced Innovation powered by the Mathematics Platform and Japanese Society of Applied Statistics. This workshop and the editing process for this volume were financially supported by the KAKENHI grant (18H00835). The editors very much appreciate the generous support from these sponsors. Also their thanks are due to people who cooperated in bringing this work to fruition. Particularly, the reliability and accuracy of the information herein were made possible due to the generosity of many esteemed anonymous reviewers.

Last but not least, the editors would like to express their sincere gratitude to Professor Sibuya for his decades of mentorship to them.

Tokyo, Japan   Nobuaki Hoshino
August 2020   Shuhei Mano
                                                                                       Takaaki Shimura

Contents

Chapter 1
Gibbs Base Random Partitions

Masaaki Sibuya

Abstract As a typical family of random partitions on $\mathcal{P}_{n,k}$, the set of partitions of n into k parts, the conditional distribution of Pitman's random partition, termed as the Gibbs base random partition, GBRP (α), is investigated. The set $\mathcal{P}_{n,k}$ is a lattice with respect to majorization partial order with unique minimum and maximum, and GBRP (α) has TP2 with respect to this order. The main purpose of this paper is to study such a family of random partitions and the inference on its parameter.

Keywords A-hypergeometric distributions · Distribution functions on a majorization order poset · Ewens-Pitman sampling formula · Part-block representation of partitions · Random partitions of n into k parts · Total order using TP2

1.1 Introduction

The Ewens-Pitman sampling formula, EPSF (θ, α), recalled in §1.2.1, is now the representative of parametric families of random partitions, which is applied successfully to the wide range of fields. Its special case, the Ewens sampling formula, EPSF $(\theta, 0)$, is its origin and characterized in many ways, because of specific features. The parameter space of EPSF is divided into three regions, (i) the main region, $0 < \alpha < 1$, $0 < \theta + \alpha < \infty$; (ii) the degenerate region, $-\infty < \alpha < 0$, $m = -\theta/\alpha = 2, 3, \ldots$; (iii) the Ewens sampling formula, $\alpha = 0$, the border between (i) and (ii). Hence, it is natural to question whether an observed partition comes from the Ewens sampling formula or not. This is the main motivation of the present paper. For this purpose, note that given the number of parts, the conditional partition is independent of θ. We refer the conditional partition as *the Gibbs base random partition*, GBRP (α), which is formally defined in §1.2.3. To obtain the better understanding

M. Sibuya (✉)
Keio University, 3-14-1 Hiyoshi, Kohoku-ku, Yokohama-shi, Kanagawa 223-8522, Japan
e-mail: sibuyam@1986.jukuin.keio.ac.jp

© The Author(s), under exclusive license to Springer Nature Singapore Pte Ltd. 2020   1
N. Hoshino et al. (eds.), *Pioneering Works on Distribution Theory*,
JSS Research Series in Statistics,
https://doi.org/10.1007/978-981-15-9663-6_1

of EPSF, the study of GBRP may be helpful as a complement of EPSF $(\theta, 0)$. This is the other motivation for the present research.

The set of all partitions of n into k parts, $\mathcal{P}_{n,k}$, or its union $\mathcal{P}_n = \bigcup_{k=1}^n \mathcal{P}_{n,k}$ is partially ordered by majorization, and they are lattices with unique maximum and minimum partitions. Along with the majorization order, the probability mass function (pmf) of partitions of GBRP (α) is TP2, and it changes smoothly on $(-\infty, 1)$. Unexpectedly, $\alpha = 0$ is not a special parameter value in GBRP (α) except for that some expressions are simple at $\alpha = 0$.

Concerning EPSF, the recent survey by Crane [3] with discussions is highly recommended. See also text books by Feng [5] and Pitman [15] and the preceding survey by Tavaré and Ewens [21]. Random partitions on $\mathcal{P}_{n,k}$ are called microcanonical distributions in physics, and the specific role of GBRP within the Gibbs random partitions is shown by Gnedin and Pitman [8].

In the statistical theory, the concept of partial order is not popular. Among some of the works dealing with partially ordered sets (*posets*), a typical subject is the measure of agreement between two types of posets; see [16]. Another typical subject is the concept of *partial quantile* defining the quantile of the random variable in the general poset; see [2]. Since we are concerned with the parametric family of probability distributions on finite posets, our approach is different from theirs.

Although the current topic on random partitions is *Bayesian nonparametric statistics* (see, e.g., [6] or [7]), the present paper is classical and elementary, and numerical calculations and visualizations are devised. A more profound study, regarding GBRP (α) as an A-hypergeometric distribution defined by Takayama et al. [20] was done by Mano [13].

In the remaining part of introduction, the partition of a number and the majorization order of partitions are reviewed in short. In Sect. 1.2, the properties of GBRP (α), including the parameter estimation related to [13], are studied. In Sect. 1.3 the tests of hypotheses on α are studied, and problems in numerical computations are discussed. In Appendix, lower adjacency in GBRP, the majorization order of integer vectors, and partial order in GBRP are briefly discussed.

1.1.1 Partitions of a Number

Let $\mathbb{N}$ denote the set of positive integers. If $n \in \mathbb{N}$ and $c_1 + \cdots + c_k = n$, $1 \leq i \leq k$, the set $c = \{c_1, \ldots, c_k\}$, $1 \leq k \leq n$ is a partition of n, expressed as $c \vdash n$. The elements c_i's are called parts (clusters, blocks, or species). If $|c| := c_1 + \cdots + c_k$, $|c| = n$ is equivalent to $c \vdash n$. Its three expressions are specified as follows. The first expression is descending order statistics (dos):

$$(c_{[1]}, c_{[2]}, \ldots, c_{[k]}), \quad c_{[1]} \geq \cdots \geq c_{[k]} \quad \text{or simply} \quad (c_1, \ldots, c_k) \downarrow .$$

Its visualization, the Ferrers diagram (or the Young diagram), is a layout of n squares in k flush left rows with lengths $c_{[1]}, \ldots, c_{[k]}$. The second expression is the size index (the terminology proposed by Sibuya [18] for the size of parts):

$$s = (s_1, \ldots, s_n), \quad s_j := \sum_{i=1}^{n} \mathbb{I}[c_i = j], \quad \mathbb{I}[\text{TRUE}] = 1, \mathbb{I}[\text{FALSE}] = 0,$$

$$\sum_{j=1}^{n} s_j = k, \sum_{j=1}^{n} j s_j = n.$$

The third one is a combination of the above two or the list of part-blocks:

$$((j_i, s_{j_i}), \quad j_1 > \cdots > j_b > 0),$$

which is the reversed sequence of the classical expression $1^{s_1} 2^{s_2} \cdots$ and usual observation records.

For example, the three expressions of an example partition, and a corresponding Ferrers diagram with part-block frames are shown below.

$$n = 26, \ k = 10, \ b = 4$$
$$(6, 4, 4, 4, 2, 2, 1, 1, 1, 1, 0, \ldots, 0)$$
$$(4, 2, 0, 3, 0, 1, 0, \ldots, 0)$$
$$\begin{bmatrix} 6 & 4 & 2 & 1 \\ 1 & 3 & 2 & 4 \end{bmatrix}$$

These three expressions will be liberally used to denote a partition. Hereinafter, letters n, k, and b are used exclusively in the above sense throughout the present paper. The set of partitions of n into k parts is denoted by $\mathcal{P}_{n,k}$. Further, $\mathcal{P}_n := \bigcup_{k=1}^{n} \mathcal{P}_{n,k}$, $\mathcal{P} := \bigcup_{n=1}^{\infty} \mathcal{P}_n$. $|\mathcal{P}_n|$ and $|\mathcal{P}_{n,k}|$ are called *partition numbers*.

Proposition 1.1 *The minimum number of part-blocks in $\mathcal{P}_{n,k}$ is $b = 1$ or 2. If n mod $k = 0$, there is a partition of one block: $((n/k)^k)$. Otherwise, there are a number of partitions of $b = 2$.*
The maximum number b_{max} of part-blocks in $\mathcal{P}_{n,k}$ is as follows.
If $\binom{k+1}{2} \leq n$, then $b_{max} = k$. Otherwise, b_{max} is the solution of $\binom{b}{2} \leq n - k < \binom{b+1}{2}$.

Proof (of the maximum part) It is easy to see, $b = k$ iff $\binom{k+1}{2} \leq n$. Discard the leftmost column of the Ferrers diagram, that is, subtract one from all parts. Then $n^* := n - k$ (squares) must be allocated into $k(s_1) := k - s_1$ parts. The above inequlity, being applied to the pair $(n^*, k(s_1))$, implies that, if $\binom{k(s_1)+1}{2} \leq n^*$, then $b_{max}(s_1) = k(s_1) + 1$. The minimum s_1 satisfying $\binom{k(s_1)+1}{2} \leq n^*$, say s_1^*, provides $b_{max} = b_{max}(s_1^*) = k(s_1^*)$. In other words, $\binom{b_{max}}{2} \leq n^* < \binom{b_{max}+1}{2}$. □

Figure 1.1 illustrates the maximum number of part-blocks of partitions in $\mathcal{P}_{n,k}$, $n = 100$. The maximum number in $\mathcal{P}_n$ is approximately $\sqrt{2n}$. Proposition 1.1 is intended

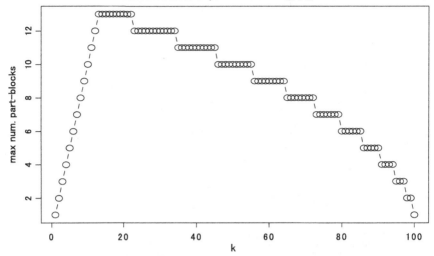

Fig. 1.1 Maximum number of part-blocks of partitions in $\mathcal{P}_{100,k}$

to check the efficiency of part-block-wise operations compared to part-wise operations when n and k are large. An example is the function genpart(n, k, m) given in the next paragraph. The depth of recursive calls is substantially reduced.

The partitions of $\mathcal{P}_n$ are totally (or linearly) ordered by the reverse lexicographical order. For example, if $n = 6$, in dos expression,

$$6 \mid 51 \mid 42, \ 41^2 \mid 3^2, \ 321, \ 31^3 \mid 2^3, \ 2^21^2, \ 21^4 \mid 1^6.$$

In this paper, the priority is given to the order of k, and within $\mathcal{P}_{n,k}$, partitions are ordered by the reverse lexicographical order (*rlg order*). For example, in $\mathcal{P}_6$,

$$6 \mid 51, \ 42, \ 3^2 \mid 41^2, \ 321, \ 2^3 \mid 31^3, \ 2^21^2 \mid 21^4 \mid 1^6. \tag{1.1}$$

A way for generating all partitions of $\mathcal{P}_{n,k}$ is the recursive call of the function genpart(n, k, m), with $m \geq n - k + 1$. The function generates all partitions of $\mathcal{P}_{n,k}$ whose maximum size of parts is restricted by m. The result is ordered like (1.1). A useful total order, which is specific to GBRP, is introduced in §1.3.1.

S/R-function 'genpart'

output matrix of
genpart(12, 5, m), $8 \le m$.

```
genpart<-function(n,k,m){
    if(n==k) return(matrix(1, n, 1))
    if(k==1) return(matrix(n, 1, 1))
    mxj <- min(m,n-k+1)
    mnj <- ceiling(n/k)
    mm <- NULL
    for(j in mxj:mnj){
        ww <- Recall(n-j,k-1,j)
        vv <- rbind(j,ww)
        mm <- cbind(mm,vv)
    }
    return(mm)
}
```

$$\begin{pmatrix} 8 & 7 & 6 & 6 & 5 & 5 & 5 & 4 & 4 & 4 & 4 & 3 & 3 \\ 1 & 2 & 3 & 2 & 4 & 3 & 2 & 4 & 3 & 3 & 2 & 3 & 3 \\ 1 & 1 & 1 & 2 & 1 & 2 & 2 & 2 & 3 & 2 & 2 & 3 & 2 \\ 1 & 1 & 1 & 1 & 1 & 1 & 2 & 1 & 1 & 2 & 2 & 2 & 2 \\ 1 & 1 & 1 & 1 & 1 & 1 & 1 & 1 & 1 & 1 & 2 & 1 & 2 \end{pmatrix}$$

A well-known partial order ($\le$) for $\lambda, \mu \in \mathcal{P}$, $\lambda = (\lambda_1, \lambda_2, \dots) \downarrow$, $\mu = (\mu_1, \mu_2, \dots) \downarrow$ is defined by

$$\lambda_i \le \mu_i, \ i = 1, 2, \dots. \iff \lambda \le \mu \ \text{(or } \mu \ge \lambda).$$

Note that $\lambda \le \mu$ & $\lambda \ne \mu$ implies $|\lambda| < |\mu|$, $|\lambda| = \lambda_1 + \lambda_2 + \cdots$. Geometrically, the Ferrers diagram of λ is included in that of μ.

For any two elements λ and μ of the partially ordered set (POS) $\mathcal{Y} := (\mathcal{P}, \le)$, two binary operators $\vee$ and $\wedge$ are defined as follows, regarding them as the infinite sequence with zeros tail.

$$\lambda \vee \mu := (\max(\lambda_1, \mu_1), \max(\lambda_2, \mu_2), \dots) \ge \lambda, \mu,$$
$$\lambda \wedge \mu := (\min(\lambda_1, \mu_1), \min(\lambda_2, \mu_2), \dots) \le \lambda, \mu.$$

The operator $\vee$ denotes union, and $\wedge$ denotes meet, of the Ferrers diagrams. For any $\lambda, \mu \in \mathcal{Y}$, there exist uniquely $\lambda \wedge \mu$ and $\lambda \vee \mu$. Hence $\mathcal{Y}$ is a lattice, and called *the Young lattice*.

1.1.2 Majorization Order of Partitions

The Young lattice is strong and orders partitions of different n, whereas we are interested in weaker order for comparison within $\mathcal{P}_n$ or $\mathcal{P}_{n,k}$. The *majorization order* ($\prec$) (or dominance order) for $\lambda = (\lambda_1, \lambda_2, \dots) \downarrow$ and $\mu = (\mu_1, \mu_2, \dots) \downarrow$ in $\mathcal{P}_n$ is defined by

$$\sum_i^k \lambda_i \le \sum_i^k \mu_i, \ \ k = 1, \dots, n. \iff \lambda \prec \mu \ \text{(or } \mu \succ \lambda).$$

Example

$\mathcal{P}_6$, $|\mathcal{P}_6| = 11$. The layout is increasing rightward and upward.

$$
\begin{array}{cccccc}
 & & & & \prec 51 & \prec 6 \\
 & & & \prec 41^2 & \prec 42 & \\
 & & \prec 31^3 & \prec 321 & \prec 3^2 & \\
1^6 & \prec 21^4 & \prec 2^2 1^2 & \prec 2^3 & &
\end{array}
$$

Note that $2^3 \not\prec 31^3$ and $3^2 \not\prec 41^2$. This example is too simple, and the layout is misleading. For larger n, even restricted to $P_{n,k}$, partitions cannot be arranged on a plane grid.

Binary operators $\vee$ and $\wedge$ are defined within $(\mathcal{P}_{n,k}, \prec)$, that is;

Proposition 1.2 *For any $\lambda, \mu \in \mathcal{P}_{n,k}$, there exist unique $\lambda \vee \mu$ and $\lambda \wedge \mu$ within $\mathcal{P}_{n,k}$. Hence, $(\mathcal{P}_{n,k}, \prec)$ is a lattice.*

Proof If $k = 1, 2$ or $k = n - 1, n$, the proposition is trivial. In general, it is proven constructively. Let the cumulative sum of dos expressions be denoted as $\lambda_j^+ = \sum_{i=1}^{j} \lambda_i$, $\mu_j^+ = \sum_{i=1}^{j} \mu_i$, $1 \le j \le k$, then λ_j^+ and μ_j^+ are increasing and concave functions in j, and $c_j^* := \min(\lambda_j^+, \mu_j^+)$, $1 \le j \le k$, is also increasing and concave. Hence, $c_j := c_j^* - c_{j-1}^*$ ($c_0^* = 0$) is positive, decreasing, $c \in \mathcal{P}_{n,k}$, and $c = \lambda \wedge \mu$.

Next, let $d_j^* := \max(\lambda_j^+, \mu_j^+)$, $d_j^o = d_j^* - d_{j-1}^*$ ($d_0^* = 0$), and $d^o := (d_1^o, \ldots, d_n^o)$ is not always decreasing, so let d be the rearrangement of d^o in decreasing order. Since d is the minimum of concave majorants of λ^+ and μ^+, $d = \lambda \vee \mu$. □

Since $\lambda_j = \lambda_j^+ - \lambda_{j+1}^+$, $j = 1, 2, \ldots$, $\lambda^+ := (\lambda_1^+, \lambda_2^+, \ldots)$, the cumulative sum of dos, is another expression of partitions.

There exist unique lower and upper limits in $\mathcal{P}_{n,k}$:

$$
\min\{v \in \mathcal{P}_{n,k}\} = (\overbrace{\lceil n/k \rceil, \ldots}^{n \bmod k}, \overbrace{\lfloor n/k \rfloor, \ldots}^{k-n \bmod k}), \quad \text{and} \quad \max\{v \in \mathcal{P}_{n,k}\} = (n - k + 1, 1^{k-1}).
$$

Note that both limits have two or one part-blocks.

Proposition 1.2 is extended to the fact that $(\mathcal{P}_n, \prec)$ is a lattice. Further, in $\mathcal{P}_{n;k_L:k_U} := \bigcup_{k_L \le k \le k_U} \mathcal{P}_{n,k}$, $1 \le k_L \le k_U \le n$, $\min\{v \in \mathcal{P}_{n,k_L}\}$ and $\max\{v \in \mathcal{P}_{n,k_U}\}$ are lower and upper limits, respectively. $\mathcal{P}_n$ is its special case where $\{1^n\}$ and $\{n\}$ are limits.

The theory of the majorization order is an important and established principle. All aspects of this theory are covered in Marshall et al. [14]. See a paragraph in Appendix.

Proposition 1.3 *The majorization order $\lambda \prec \mu$ implies both lexicographical order and reverse lexicographical order (rlg), $\lambda \le_{\text{rlg}} \mu$, (1.1), in $\mathcal{P}_{n,k}$ and $\mathcal{P}_n$.*

Proof   If $\lambda \prec \mu$, $\sum_{i=1}^{j} \lambda_i \leq \sum_{i=1}^{j} \mu_i$, $1 \leq j \leq k$, which means that for some m, $1 \leq m < k$, $\lambda_i \leq \mu_i$, $i = 1, \ldots, m$. Hence $\lambda \leq_{\mathrm{rlg}} \mu$.   □

1.1.3 Adjacency

To see finer structures in POS $(\mathcal{P}_n, \prec)$, we define a relation $\lambda \lessdot \mu$, λ *is smaller than and adjacent to* μ (or $\mu \gtrdot \lambda$), if

$$\lambda \prec \nu \prec \mu \Leftrightarrow \lambda = \nu \text{ or } \nu = \mu.$$

For examining the majorization order structure in $\mathcal{P}_{n,k}$ (or $\mathcal{P}_n$), we need to identify the upper or larger adjacent partitions of each partition of $\mathcal{P}_{n,k}$ (or $\mathcal{P}_n$). For this purpose, make a part-block less equal; take an element of a part-block, and move it to another part of the same part-block, otherwise, if the part-block consists of a single part, move it to a larger part. According to the choice of a part-block, the change can be one of the following two moves:

(A)   Move to an upper partition of $\mathcal{P}_{n,k}$.
(B)   Move from the partitions of $\mathcal{P}_{n,k}$ to a partition of $\mathcal{P}_{n,k-1}$.

We can consider a dual way: (C) Move to a lower partition of $\mathcal{P}_{n,k}$, and (D) Move from the partitions of $\mathcal{P}_{n,k}$ to a partition of $\mathcal{P}_{n,k+1}$. These moves are summarized in Table 1.10 in Appendix.
Further details are as follows. The remarked part-block will be denoted by (j_i, s_{j_i}).

A1. Condition $j_i > 1$ and $s_{j_i} > 1$.
Take an element of a part-block (j_i, s_{j_i}) such that $j_i > 1$ and $s_{j_i} > 1$, and put it into another part of the same part-block. The result of this move is classified further as follows:

a. If $i = 1$ and $j_1 - j_2 > 1$,

$$(j_1 + 1, 1)(j_1, s_{j_1} - 2)(j_1 - 1, 1) * * * .$$

b. If $i > 1$ and $j_{i-1} - j_i > 1$,

$$\cdots (j_i + 1, 1)(j_i, s_{j_i} - 2)(j_i - 1, 1) * * * .$$

c. If $i > 1$ and $j_{i-1} - j_i = 1$,

$$\cdots (j_{i-1}, s_{j_{i-1}} + 1)(j_i, s_{j_i} - 2)(j_i - 1, 1) * * * .$$

In these results, if $s_{j_i} = 2$, the middle part-block disappears. The symbol $\cdots$ denotes unchanged part-blocks, and $* * *$ denotes generic part-blocks affected by $(j_i - 1, 1)$ as explained later.

In **A1**, the case of $j_i > 1$ and $s_{j_i} = 1$ is excluded. In this situation, move an element of $(j_i, s_{j_i} = 1)$ to the upper block $(j_{i-1}, s_{j_{i-1}})$, and the result will be

$$\cdots (j_{i-1} + 1, 1)(j_{i-1}, s_{j_{i-1}} - 1)(j_i - 1, 1) * * * .$$

However, this result can be obtained through two moves. First, apply move A1 to $(j_{i-1}, s_{j_{i-1}})$ to obtain

$$\cdots (j_{i-1} + 1, 1)(j_{i-1}, s_{j_{i-1}} - 2)(j_i, 1) * * * .$$

Second, move an element of $(j_i, 1)$ to the upper part-block. Hence, the above move does not result in adjacent partition, provided that $s_{j_{i-1}} > 1$, which is a necessary condition in the above two move argument. Hence, the last possible situation is

A2. Condition $j_i > 1$, $s_{j_i} = 1$ and $s_{j_{i-1}} = 1$.

In this specific situation, an element of $(j_i, 1)$ can be moved into $(j_{i-1}, 1)$.

a. If $i = 2$, the first two part-blocks are involved, and the result is

$$(j_1 + 1, 1)(j_2 - 1, 1) * * * .$$

b. Generally, if $i > 2$ and $j_{i-2} - j_{i-1} > 1$, two part-blocks are involved, and the result is similar to the case a.

$$\cdots (j_{i-1} + 1, 1)(j_i - 1, 1) * * * .$$

c. Finally, if $i > 2$ and $j_{i-2} - j_{i-1} = 1$, three part-blocks are involved, and the result is

$$\cdots (j_{i-2}, s_{j_{i-2}} + 1)(j_i - 1, 1) * * * .$$

Throughout A1 and A2, the last part-block and $* * *$ should be remarked. If (j_i, s_{j_i}) is the last part-block, namely, $i = b$, the new last part-block is $(j_i - 1, 1)$. If $j_i - j_{i+1} > 1$, the succeeding part-blocks are unchanged, and if $j_i - j_{i+1} = 1$, these will be $(j_i - 1, s_{j_i-1} + 1) \cdots$.

Proposition 1.4 *The change of a partition in $\mathcal{P}_{n,k}$ to the adjacent upper partitions in $\mathcal{P}_{n,k}$, by the move of an element of a part-block, is summarized in part A of Table 1.2.*

See, as an example, Table 1.1 of adjacency in $\mathcal{P}_{16,6}$.

Let $\tau = (\tau_1, \ldots, \tau_n)$, $\tau \vdash n$, be the size index expression of τ. The result of Move A1 and A2 of τ is simple in this expression:

$$\text{A1; } (\ldots, \tau_{i-1} + 1, \tau_i - 2, \tau_{i+1} + 1, \ldots), \quad 1 < \tau_i, \; 1 < i < \lfloor n/2 \rfloor, \tag{1.2}$$

$$\text{A2; } (\ldots, \tau_{i-1} + 1, \tau_i - 1, \tau_j - 1, \tau_{j+1} + 1, \ldots), \; 1 < i < j \le n - k, \, j - i = 1, \tau_i = \tau_j = 1. \tag{1.3}$$

If $j_i = 1$, which is the case excluded from case A, a part of the last part-block disappears, and the partition changes from $\mathcal{P}_{n,k}$ to $\mathcal{P}_{n,k-1}$. The details are similar to

Table 1.1 Adjacency in $\mathcal{P}_{16,6}$, $|\mathcal{P}_{16,6}| = 35$. There are 52 adjacent pairs. Partitions are arranged from the lower to the upper. In 1(136) etc., the parenthesized is the rlg order in $\mathcal{P}_{16}$

36-rlg	1(136)	2	3	4	5	6	7	8	9	10	11	12(125)
DOS expression	3^42^2	3^51	4^32^3	4^321	4^22^4	4^232^21	$4^23^21^2$	4^321^2	532^4	53^22^21	53^31^2	542^31
upper adjacent partitions	2 3	4	4 5	6	6 9	7 10	8 11	13	10 17	11 12	13	13 18
lower adjacent partitions		1	1	3 2	3	5 4	6	7	5	9 6	10 6	10 7

36-rlg	13(124)	14	15	16	17	18	19	20	21	22	23	24(113)
DOS expression	54321^2	54^21^3	$5^22^21^2$	5^231^3	62^5	632^31	63^221^2	642^21^2	6431^3	6521^3	6^21^4	72^41
upper adjacent partitions	14 15 19	16	16 20	21	18	19 24	20	21 25	22 26	23 27	28	25
lower adjacent partitions	12 11 8	13	13	15 14	9	17 12	18 13	19 15	20 16	21	22	18

36-rlg	25(112)	26	27	28	29	30	31	32	33	34	35(102)
DOS expression	732^21^2	73^21^3	7421^3	751^4	82^31^2	8321^3	841^4	92^21^3	931^4	$10\,21^4$	$11\,1^5$
upper adjacent partitions	26 29	27	28 30	31	30	31 32	33	33	34	35	
lower adjacent partitions	24 20	25 21	26 22	27 23	25	29 27	30 28	30	32 31	33	34

case A. Remark that $j_i = 1$ means that the part-block is the last one, so that $i = b$, that is, the number of part-blocks of the partition.

B1. Condition $j_i = 1$ and $s_{j_i} > 1$.
Take an element of the part-block $(1, s_1)$ such that $s_1 > 1$, and put it into another part of the same part-block.

a. If $i = 1$, the partition has only one part-block that $k = n$ and changes to $(2, 1)(1, n - 2) \in \mathcal{P}_{n,n-1}$.
b. If $b > 1$ and $j_{b-1} > 2$, the last part-block $(1, s_1)$ changes to $\cdots (2, 1)(1, s_1 - 2)$.
c. If $b > 1$ and $j_{b-1} = 2$, it follows that $\cdots (2, s_2 + 1)(1, s_1 - 2)$.

B2. Condition $j_b = 1$, $s_{j_b} = 1$ and $s_{j_{b-1}} = 1$.
Note that if $s_{j_b} = 1$, as in Case A2, adjacent upper partitions are possible only if $s_{j_{b-1}} = 1$. The last part-block is $(1, 1)$, and its element is moved to the upper adjacent part-blocks.

a. If $i = b = 2$ and $(j_1, s_{j_1}) = (n - 1, 1)$, the new partition is

$$(n, 1) \in \mathcal{P}_{n,1}.$$

b. If $i = b > 2$ and $j_{b-2} - j_{b-1} > 1$, the new last part-block is

$$\cdots (j_{b-1} + 1, 1).$$

c. If $i = b > 2$ and $j_{b-2} - j_{b-1} = 1$, the new last part-block is

$$\cdots (j_{b-2}, s_{j_{b-2}} + 1).$$

Table 1.2 Classification of the *upward* transforms from a partition $\{(j_i, s_{j_i}), \ 1 \le i \le b\}$. The type of part-blocks, from which one element is to be taken out, is classified.   A: within $\mathcal{P}_{n,k}$, B: from $\mathcal{P}_{n,k}$ to $\mathcal{P}_{n,k-1}$

$j_i \setminus s_{j_i}$	$s_{j_i} > 1$	$s_{j_i} = 1$
A: $j_i > 1$	A1:	A2: $(i > 1, s_{j-1} = 1)$
B: $j_i = 1$	B1: $(j_b = 1)$	B2: $(i > 1, s_{j-1} = 1)$

Change of part-blocks in each transform	
A1	A2 (if $s_{j_{i-1}} = 1$)
$i = 1$	$i = 2$
$(j_1 + 1, 1)(j_1, s_{j_1} - 2)(j_1 - 1, 1) * **$	$(j_1 + 1, 1)(j_2 - 1, 1) * **$
$i > 1, j_{i-1} - j_i > 1$	$i > 2, j_{i-2} - j_{i-1} > 1$
$\cdots (j_i + 1, 1)(j_i, s_{j_i} - 2)(j_i - 1, 1) * **$	$\cdots (j_{i-1} + 1, 1)(j_i - 1, 1) * **$
$i > 1, j_{i-1} - j_i = 1$	$i > 2, j_{i-2} - j_{i-1} = 1$
$\cdots (j_{i-1}, s_{j_{i-1}} + 1)(j_i, s_{j_i} - 2)(j_i - 1, 1) * **$	$\cdots (j_{i-2}, s_{j_{i-2}} + 1)(j_i - 1, 1) * **$

Change of lower part-blocks. A1 and A2	
$i = b$	$\cdots (j_i - 1, 1).$
$j_i - j_{i+1} > 1 \ (i < b) \cdots (j_i - 1, 1) \cdots$	
$j_i - j_{i+1} = 1 \ (i < b) \cdots (j_{i+1} = j_i - 2, s_{j_{i+1}} + 1) \cdots$	

Change of part-blocks in each transform	
B1 $j_i = 1, s_1 > 1$	B2 $j_i = 1, s_1 = 1$
$i = b = 1$ $(2, 1)(1, n - 2).$	$i = b = 2$ $(n, 1).$
$b > 1, j_{b-1} > 2 \cdots (2, 1)(1, s_1 - 2).$	$b > 2, j_{b-2} - j_{b-1} > 1 \cdots (j_{b-1} + 1, 1).$
$b > 1, j_{b-1} = 2 \cdots (2, s_2 + 1)(1, s_1 - 2).$	$b > 2, j_{b-2} - j_{b-1} = 1 \cdots (j_{b-2}, s_{j_{b-2}} + 1).$

Proposition 1.5 *The change of a partition in $\mathcal{P}_{n,k}$ to the adjacent upper partitions in $\mathcal{P}_{n,k-1}$, by the move of an element of a part-block, is summarized in part B of Table 1.2.*

In the size index expression of τ, the results of Move B1 and B2 of τ are as follows:

B1; $(\tau_1 - 2, \tau_2 + 1, \ldots), \quad 1 < \tau_1.$

B2; $(0. \ldots, 0, \tau_{j+1} + 1, \ldots), \quad \tau_1 = \tau_j = 1, j > 1$, and $\tau_i = 0, 1 < i < j.$

Its implication is out of the scope of the present paper.

Upper set generation

Given $\lambda \in \mathcal{P}_{n,k}$, we construct its upper set $\mathcal{U}(\lambda) := \{v \in \mathcal{P}_{n,k}; \lambda \prec v\}$. All partitions will be expressed in dos, or equivalently, in a cumulative sum of dos sequence. Let

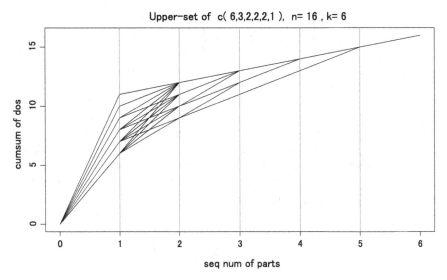

Fig. 1.2 Cumulative sum expression of partitions of $\mathcal{U}(\lambda)$, $\lambda = 632^31 \in \mathcal{P}_{16,6}$

μ denote the maximum ($\prec$) partition of $\mathcal{U}(\lambda)$, or of $\mathcal{P}_{n,k}$. That is, $\mu = (n - k + 1, 1^{k-1})$, and

$$\mathcal{U}(\lambda) = \{v; \lambda \prec v \prec \mu\} = \{v^+; \lambda^+ \le v^+ \le \mu^+\},$$

where $v_j^+ = \sum_{i=1}^{j} v_i$, $1 \le j \le k$, and $v^+ \le \mu^+$ represents the component-wise comparison. The cumulative sum of a partition of $\mathcal{U}(\lambda)$ is shown as an increasing concave polygonal line limited by those of λ and μ. Figure 1.2 illustrates (j, v_j^+), $1 \le j \le k$, of $v \in \mathcal{U}(\lambda)$ of an example that $\lambda = 632^31$, where $\mu = 11.1^5$ and $|\mathcal{U}(\lambda)| = 18$ in $\mathcal{P}_{16,6}$. See also Table 1.1 of adjacency in $\mathcal{P}_{16,6}$.

First, we describe the outline of the algorithm for generating an upper set. In Fig. 1.2, all lines connecting λ and μ are $v^+ = (v_1^+, \ldots, v_k^+)$ such that $\lambda_j^+ \le v_j^+ \le \mu_j^+$, $1 \le j \le k$. From these candidates, we select the increasing and concave lines:

$$v_{j-1} < v_j, 1 < j \le k \text{ and } v_{j-2} - 2v_{j-1} + v_j \le 0.$$

The problem here is how to avoid redundant works and save memory. The output of the algorithm is a $k \times |\mathcal{U}(\lambda)|$ matrix, whose columns are cumulative sums v^+ of $v \in \mathcal{U}(\lambda)$, combined in rlg order. We construct the matrix starting from its bottom, 1×1 matrix $[n]$, moving up row by row and thereby, increasing the number of columns.

Algorithm for $\mathcal{U}(\lambda)$. If λ has a part-block 1^c, the last c segments of broken lines are reduced to one straight line: $\lambda_j^+ = \mu_j^+$, $m \le j \le k$, $m = k - c$. We start from a matrix M_{m-1}:

$$M_m = \begin{pmatrix} \lambda_m^+ \\ \vdots \\ \vdots \\ \lambda_k^+ \end{pmatrix}, \qquad M_{m-1} = \begin{pmatrix} \lambda_{m-1}^+ & \cdots & \mu_{m-1}^+ \\ \lambda_m^+ & \cdots & \lambda_m^+ \\ \cdots & \cdots & \cdots \\ \lambda_k^+ & \cdots & \lambda_k^+ \end{pmatrix},$$

$$\lambda_m^+ < \cdots < \lambda_k^+, \text{ and } \lambda_{m-1}^+ < \cdots < \mu_{m-1}^+.$$

The procedure for constructing M_{j-1} from M_j, $j = m-1, \ldots, 1$, is as follows.

The candidates to append to subsequences (columns) are $\underline{\ell} = (\ell_1, \ell_2, \ldots) := (\lambda_{j-1}^+, \ldots, \mu_{j-1}^+)$. Append the vector of ℓ_1 to $M_j^{(1)} := M_j$ as the top row; check the new columns whether they are strictly increasing and concave, and delete inadequate columns. Thus, the columns of $M_j^{(1)}$ are reduced to, say, $M_j^{(2)}$. Next, append the vector of ℓ_2 to $M_j^{(2)}$ as the top row, and delete inadequate columns.

$$\begin{pmatrix} \ell_1 & \cdots & \ell_1 \\ & M_j^{(1)} & \end{pmatrix}, \ldots, \begin{pmatrix} \ell_i & \cdots & \ell_i \\ & M_j^{(i)} & \end{pmatrix} \cdots$$

Repeat this procedure up to $\ell_i = \mu_{j-1}^+$, and join the matrices column-wise to obtain M_{j-1}.

If M_1 is obtained, the final procedure is to append the row vector of zeros on the top of M_1, and check the concavity. Reduced M_1 is the cumulative sum expression of $\mathcal{U}(\lambda)$. □

Moving to the next ℓ_{i+1} ($\ell_i < \ell_{i+1}$), the matrix $M_j^{(i)}$ is reduced at each step for efficiency. If $(\ell_i, v_j, \ldots, v_k)$ is decreasing or convex, $(\ell_{i+1}, v_j, \ldots, v_k)$ cannot be strictly increasing or concave.

A numerical example for $\mathcal{U}(\lambda)$ of Fig. 1.2 is provided as follows:

$$n = 16, k = 6, m = 5, \lambda^+ = (6, 9, 11, 13, 15, 16), \text{ and } \mu^+ = (11, 12, 13, 14, 15, 16).$$

$$M_5 = \begin{bmatrix} 15 \\ 16 \end{bmatrix}; \qquad M_4 = \begin{bmatrix} 13 & 14 \\ 15 & 15 \\ 16 & 16 \end{bmatrix};$$

$$\begin{bmatrix} \ell_1 \\ M_3^{(1)} \end{bmatrix} = \begin{bmatrix} 11 & 11 \\ 13 & 14 \\ 15 & 15 \\ 16 & 16 \end{bmatrix}, \begin{bmatrix} \ell_2 \\ M_3^{(2)} \end{bmatrix} = \begin{bmatrix} 12 & 12 \\ 13 & 14 \\ 15 & 15 \\ 16 & 16 \end{bmatrix} \text{ delete 1st column, } \begin{bmatrix} \mu_3^+ \\ \mu_4^+ \\ M_3^{(3)} \\ \mu_5^+ \\ \mu_6^+ \end{bmatrix} = \begin{bmatrix} 13 \\ 14 \\ 15 \\ 16 \end{bmatrix}, \quad M_3 = \begin{bmatrix} 11 & 11 & 12 & 13 \\ 13 & 14 & 14 & 14 \\ 15 & 15 & 15 & 15 \\ 16 & 16 & 16 & 16 \end{bmatrix};$$

$$\begin{bmatrix} \ell_1 \\ M_2^{(1)} \end{bmatrix} = \begin{bmatrix} 9 & 9 & 9 & 9 \\ 11 & 11 & 12 & 13 \\ 13 & 14 & 14 & 14 \\ 15 & 15 & 15 & 15 \\ 16 & 16 & 16 & 16 \end{bmatrix} \text{ delete 2nd column, } \begin{bmatrix} \ell_2 \\ M_2^{(2)} \end{bmatrix} = \begin{bmatrix} 10 & 10 & 10 \\ 11 & 12 & 13 \\ 13 & 14 & 14 \\ 15 & 15 & 15 \\ 16 & 16 & 16 \end{bmatrix} \text{ delete 1st column,}$$

$$\begin{bmatrix} \ell_3 \\ M_2^{(3)} \end{bmatrix} = \begin{bmatrix} 11 & 11 \\ 12 & 13 \\ 14 & 14 \\ 15 & 15 \\ 16 & 16 \end{bmatrix} \text{ delete 1st column, } \begin{bmatrix} \mu_2^+ \\ \mu_3^+ \\ \mu_4^+ \\ M_2^{(4)} \\ \mu_5^+ \\ \mu_6^+ \end{bmatrix} = \begin{bmatrix} 12 \\ 13 \\ 14 \\ 15 \\ 16 \end{bmatrix}, \quad M_2 = \begin{bmatrix} 9 & 9 & 9 & 10 & 10 & 11 & 12 \\ 11 & 12 & 13 & 12 & 13 & 13 & 13 \\ 13 & 14 & 14 & 14 & 14 & 14 & 14 \\ 15 & 15 & 15 & 15 & 15 & 15 & 15 \\ 16 & 16 & 16 & 16 & 16 & 16 & 16 \end{bmatrix}.$$

The construction of lower sets is a conjugate procedure. If $\lambda = \min\{\nu \in \mathcal{P}_{n,k}\}$, $\mathcal{U}(\lambda) = \mathcal{P}_{n,k}$, and the procedure is another way for constructing $\mathcal{P}_{n,k}$, with the duplicate adjacency information (ν_{j-1}, ν_j).

Alternative affirmation of adjacency

The relation $\lambda \prec \mu$ can be easily checked, and the other way to affirm adjacency in $\mathcal{P}_{n,k}$ should be considered. Let $\nu^{(i)}$, $1 \le i \le p(n, k) = |\mathcal{P}_{n,k}|$, be a sequence of partitions in rlg order. Let M be a $p(n, k) \times p(n, k)$ logical matrix with the (i, j) element

$$M_{i,j} := (\nu^{(i)} \prec \nu^{(j)}) = (\nu^{(i)} \in \mathcal{L}(\nu^{(j)})) = (\nu^{(j)} \in \mathcal{U}(\nu^{(i)})).$$

Hence, T (true) elements are restricted to the lower triangle of M, and the diagonal elements are trivially T.

Here, the purpose is to construct the logical matrix M^* with the (i, j) element

$$M^*_{i,j} = (\nu^{(i)} \lessdot \nu^{(j)}), \quad i, j \in [1, p(n, k)],$$

from M. The procedure goes downwards and by rows. In the i-th row of M, $\{\nu^{(j)}; j \in [1, i-1] \,\&\, M_{i,j} = \text{T}\}$ is the upper set of $\nu^{(i)}$, and from these partitions, delete upper partitions of $\nu^{(j)}$ recorded temporarily in $M^*_{j,..}$, the j-th row of M^*. Going through $j = i - 1, \ldots, 2$, $M^*_{j,..}$ is determined accordingly. Formally, the procedure is described below:

Algorithm.

$M^*_{1,..} = (\text{F}, \ldots, \text{F}), \quad M^*_{2,..} = (\text{T}, \text{F}, \ldots, \text{F}).$
for $i = 3, \ldots, p(n, k)\big\{$
set $V = M_{i,..}$,
for $j = i - 1, \ldots, 2\{$
if $M_{i,j} = \text{T}$, for all ℓ such as $M^*_{j,\ell} = \text{T}$, set $V_\ell = \text{F}\}$
set $M^*_{i,..} = V\big\}$
□

1.2 Gibbs Base Random Partitions

1.2.1 The Ewens-Pitman Sampling Formula

The Ewens-Pitman sampling formula, $\mathsf{EPSF}\,(\theta, \alpha)$, *or simply, the Pitman random partition* is defined by

$$w(s; \theta, \alpha; n) := \mathbb{P}\{S = s\} = \frac{(\theta| - \alpha)_k}{(\theta| - 1)_n} \pi_n(s) \prod_{j=1}^{n} ((1 - \alpha| - 1)_{j-1})^{s_j}, \qquad (1.4)$$

where

$$(\theta| - \alpha)_k := \theta(\theta + \alpha) \cdots (\theta + (k - 1)\alpha),$$

$$\pi_n(s) := \frac{n!}{\prod_{j=1}^{n} s_j!(j!)^{s_j}}, \quad s \in \mathcal{P}_{n,k} \left(\text{or } \sum_{j=1}^{n} s_j = k, \sum_{j=1}^{n} j s_j = n \right), \quad 1 \le k \le n,$$

and s is the size index expression of a partition. The generalized factorial product $(x|c)$ includes the usual descending factorial product $(x)_k = (x|1)$, the ascending factorial product $(x| - 1)_k$, and the power $(x|0)_k$. Its parameter space is

$$\{(\theta, \alpha); 0 \le \alpha \le 1, \theta + \alpha \ge 0 \quad \text{or} \quad \alpha < 0, -\theta/\alpha = 1, 2, \ldots \}.$$

On the boundary that $\alpha = 1$ & $\theta > -1$, $\mathbb{P}\{S = \{n\}\} = 1$, and on the boundary that $\alpha + \theta = 0$, $\mathbb{P}\{S = \{1^n\}\} = 1$. In the following, these degenerated random partitions will be neglected.

A special case,

$$w(s; \theta, 0; n) = \frac{\theta^k}{(\theta| - 1)_n} \frac{n!}{\prod_{j=1}^{n} s_j! j^{s_j}}, \quad s \in \mathcal{P}_{n,k}, \theta > 0,$$

is *the Ewens sampling formula*. Another special case,

$$w(s; \gamma, m; n) = (m)_k \prod_{j=1}^{n} \frac{1}{s_j!} \left(\frac{\gamma + j - 1}{j} \right)^{s_j} / \binom{m\gamma + n - 1}{n},$$

$$s \in \mathcal{P}_{n,k}, \gamma = -\alpha, m = \theta/\gamma = 2, 3, \ldots,$$

is *the symmetric negative hypergeometric distribution*.
The number of parts
Let $S = (S_1, \ldots, S_n)$ be the size index of a random partition in $\mathcal{P}_n$. $K := S_1 + \cdots + S_n$ is the number of parts, and $f_n(k; \theta, \alpha) := \mathbb{P}\{K = k\}$, referred to as **EPSF-K** $(n; \theta, \alpha)$, satisfies the *forward equation*:

$$f_{n+1}(k; \theta, \alpha) = \frac{n - k\alpha}{\theta + n} f_n(k; \theta, \alpha) + \frac{\theta + (k - 1)\alpha}{\theta + n} f_n(k - 1; \theta, \alpha), \quad 1 \le k \le n,$$

and

$$f_n(k; \theta, \alpha) = \frac{(\theta| - \alpha)_k}{(\theta| - 1)_n} S_{n,k}^*(\alpha), \quad 1 \le k \le n, \quad S_{n,k}^*(\alpha) := S_{n,k}(-1, -\alpha, 0), \quad (1.5)$$

where $S_{n,k}(a, b, c)$ is the exponential generalized Stirling numbers; see, e.g., [19]. See also the discussion following the definition of **GBRP** in §1.2.3.

For the Ewens sampling formula, K is a sufficient statistics, and

$$f_n(k; \theta, 0) = \mathbb{P}\{K = k\} = \frac{\theta^k}{(\theta| - 1)_n} S_{n,k}^*(0) = \begin{bmatrix} n \\ k \end{bmatrix} \frac{\theta^k}{(\theta| - 1)_n}, \quad 1 \le k \le n,$$

$$S_{n,k}^*(0) = \begin{bmatrix} n \\ k \end{bmatrix} : \quad \text{the unsigned Stirling numbers of the first kind.}$$

$$E(K) = \theta \left(\psi(\theta + n) - \psi(\theta) \right) = \theta \sum_{j=1}^{n} \frac{1}{\theta + j - 1}, \quad \psi(z) := \frac{d}{dz} \ln \Gamma(z).$$

Conditional distribution on $\mathcal{P}_{n,k}$
The conditional probability,

$$\mathbb{P}\{S = s | S \in \mathcal{P}_{n,k}\} = \mathbb{P}\{S = s | K = k\},$$

the ratio of (1.4) to (1.5), is independent of θ and equals to (1.6), *the Gibbs base random partition* defined in §1.2.3.

1.2.2 Probability Measure on $\mathcal{P}_{n,k}$

Stochastic orders on lattice

Three elemental stochastic orders, *usual, hazard rate (reverse hazard rate)*, and *likelihood ratio*, and their implications are well known. See, e.g., [17]. Here, they are described in terms of the discrete partially ordered set $(\mathcal{X}, \prec)$ with the lattice structure, with unique minimum and maximum. Let F_X, G_X, and f_X denote the distribution function (df), the survival function (sf), and the pmf, respectively, of a random quantity X on a lattice $(\mathcal{X}, \prec)$. Those of Y are similarly denoted.

The *usual stochastic order*, $X \le_{st} Y$, is defined by

$$F_X(v) \ge F_Y(v), \; \forall v \in (\mathcal{X}, \prec), \; \text{or} \; G_X(v) \le G_Y(v), \; \forall v \in (\mathcal{X}, \prec).$$

The *hazard rate order*, $X \le_{hr} Y$, is defined by

$$f_X(v)/G_X(v) \ge f_Y(v)/G_Y(v), \; \forall v \in (\mathcal{X}, \prec), \; \text{or} \; G_Y(v)/G_X(v) \; \uparrow \; (v \uparrow \text{in } (\mathcal{X}, \prec)).$$

Its dual order, *reverse hazard rate order*, $X \le_{rh} Y$, is defined by

$$f_X(v)/F_X(v) \le f_Y(v)/F_Y(v), \; \forall v \in (\mathcal{X}, \prec), \; \text{or} \; F_Y(v)/F_X(v) \; \uparrow \; (v \uparrow \text{ in } (\mathcal{X}, \prec)).$$

The *likelihood ratio order*, $X \le_{lr} Y$, is defined by

$$f_Y(v)/f_X(v) \text{ is nondecreasing in } v \in (\mathcal{X}, \prec)$$

$(c/0 = \infty$ by convention$)$. That is,

$$f_Y(\lambda)f_X(\mu) \le f_Y(\mu)f_X(\lambda), \quad \forall \lambda \prec \mu.$$

The order $X \le_{lr} Y$ implies $X \le_{hr} Y$ and $X \le_{rh} Y$, and each of these implies $X \le_{st} Y$. These comparisons are applied to the monotonicity of a parametric family of random partitions.

Distribution function and survival function on $\mathcal{P}_{n,k}$

Let $\mathcal{L}(v)$, a lower set of v, be the set of the partitions of $\mathcal{P}_{n,k}$ that are lower than v. The upper set $\mathcal{U}(v)$ is similarly defined:

$$\mathcal{L}(v) = \{\lambda \prec v; \ \lambda \in \mathcal{P}_{n,k}\}, \quad \text{and} \quad \mathcal{U}(v) = \{\mu \succ v; \ \mu \in \mathcal{P}_{n,k}\}.$$

A probability measure P on $\mathcal{P}_{n,k}$ is defined by a probability mass function $p(v)$, and

$$F(v) := P(\mathcal{L}(v)) = \sum_{\lambda \prec v} p(\lambda) \quad \text{and} \quad G(v) := P(\mathcal{U}(v)) = \sum_{\mu \succ v} p(\lambda)$$

are *the distribution function* and *the survival function*, respectively, of $P(\cdot)$ or $p(v)$. The relationship between them is

$$F(v) + G(v) - p(v) + \sum_{\omega \not\prec v \& \omega \not\succ v} p(\omega) = 1.$$

As a special case of the previous paragraph, $X \le_{lr} Y$ implies that $G_Y(v)/G_X(v)$ and $F_Y(v)/F_X(v)$ are nondecreasing.

An example is illustrated in §1.2.3. The notions can be extended to a partial order set in general.

1.2.3 Gibbs Base Random Partitions

The Gibbs base random partition is the conditional random partition of the Ewens-Pitman sampling formula, EPSF(θ, α), on $\mathcal{P}_{n,k}$. Using size index expression,

$$\varphi(s; \alpha) = \varphi(s; n, k, \alpha) := \mathbb{P}\{S = s | s \in \mathcal{P}_{n,k}\} = \frac{\pi_n(s)}{S^*_{n,k}(\alpha)} \prod_{j=1}^{n} ((1 - \alpha | - 1)_{j-1})^{s_j}.$$

$$(1.6)$$

See (1.4) for the notations. In particular,

$$\varphi(s; 0) = \frac{n!}{\prod_{j=1}^{n} s_j! j^{s_j}} \bigg/ \begin{bmatrix} n \\ k \end{bmatrix}.$$

Note that $w(s; 0, \alpha; n)$, a special case of $\text{EPSF}(\theta, \alpha)$, is a random partition with parameter α only. Hence, its relation with $\text{GBRP}(\alpha)$ is concerned.

$$w(s; 0, \alpha; n) = \lim_{\theta \to 0} w(s; \theta, \alpha; n) = \frac{(k-1)! \alpha^{k-1}}{(n-1)!} S^*_{n,k}(\alpha) \varphi(s; \alpha),$$

$$\sum_{s \in \mathcal{P}_{n,k}} w(s; 0, \alpha; n) = f_n(k; 0, \alpha) = \frac{(k-1)! \alpha^{k-1}}{(n-1)!} S^*_{n,k}(\alpha), \ 1 \le k \le n < \infty, \ \forall \alpha.$$

Hence

$$\sum_{k=1}^{n} S^*_{n,k}(\alpha)(k-1)! \alpha^{k-1} \equiv (n-1)!, \quad \forall \alpha. \tag{1.7}$$

This polynomial identity in α is another definition of $S^*_{n,k}$.

Moments of the size index

Let $S = (S_1, \ldots, S_{n-k+1})$ be the size index of $\text{GBRP}(\alpha)$ on $\mathcal{P}_{n,k}$. The joint factorial moments are

$$E\left(\prod_{j=1}^{n} (S_j)_{q_j}\right) = \frac{(n)_Q S^*_{n-Q,k-q}}{S^*_{n,k}} \prod_{j=1}^{n} \left(\frac{(1-\alpha|-1)_{j-1}}{j!}\right)^{q_j}, \tag{1.8}$$

where

$$q = \sum_{j=1}^{n} q_j, \ Q = \sum_{j=1}^{n} j q_j, \ S^*_{n,k} := S^*_{n,k}(\alpha).$$

Especially,

$$E(S_j) = \frac{(n)_j \ S^*_{n-j,k-1}}{S^*_{n,k}} \frac{(1-\alpha|-1)_{j-1}}{j!}$$

$$= \frac{1}{S^*_{n,k} \ \Gamma(1-\alpha)} \binom{n}{j} S^*_{n-j,k-1} \Gamma(j-\alpha), \quad 1 \le j \le n-k+1,$$

$$E(S_i S_j) = \frac{(n)_{i+j} \, S^*_{n-i-j,k-2} \, (1-\alpha|-1)_{i-1}(1-\alpha|-1)_{j-1}}{S^*_{n,k}} \frac{}{i! \, j!}$$

$$= \frac{n!}{(\Gamma(1-\alpha))^2 \, S^*_{n,k}} \frac{S^*_{n-i-j,k-2}}{(n-i-j)!} \frac{\Gamma(i-\alpha)}{i!} \frac{\Gamma(j-\alpha)}{j!},$$

$$1 \leq i, j \leq n-k+1, i \neq j, 3 \leq i+j \leq n-k+2,$$

$$E((S_1)_r) = \frac{(n)_r S^*_{n-r,k-r}}{S^*_{n,k}}.$$

Note that, if $i = j$, the r.h.s. of $E(S_i S_j)$ is not $E(S_j^2)$ but $E((S_j)_2)$.

1.3   Testing Statistical Hypotheses on α

1.3.1   Properties of GBRP

First, we investigate the pmf of GBRP as a function of α.

Proposition 1.6 *The pmf of the maximum ($\prec$) partition of $\mathcal{P}_{n,k}$ is increasing in α, while that of the minimum ($\prec$) is decreasing. Such a partition increasing near $\alpha = 1$ is unique in $\mathcal{P}_{n,k}$. If there is another partition with a decreasing pmf, those in its lower set are decreasing. In this sense, there is a lower set with decreasing pmfs. Hence, the maximum likelihood estimate does not exist for these tail partitions. It is anticipated that the pmf of a partition between ($\prec$) the two tail sets is unimodal.*

Proof First, the largest partition is shown to be increasing. Denote the pmf of a partition as $\varphi(v; \alpha) = \phi(v; \alpha)/S^*_{n,k}(\alpha)$. For any $v \in \mathcal{P}_{n,k}$, $\phi(v; \alpha) = c \prod_{j>1}((j - 1 - \alpha)_{j-1})^{s_j}$. This is a polynomial of α with degree $\sum_{j>1} s_j(j - 1) = n - k$; the same for all $v \in \mathcal{P}_{n,k}$. Since $(j - \alpha)/(i - \alpha), 0 < i < j$, is strictly increasing in α, and $\phi(\mu; \alpha)/\phi(v; \alpha)$, $\phi(\mu; \alpha) = c' \prod_{j=1}^{n-k}(j - \alpha)$, is strictly increasing $\forall v \in \mathcal{P}_{n,k} \setminus \mu$. Hence,

$$\frac{\sum_{v \in \mathcal{P}_{n,k}} \phi(v; \alpha)}{\phi(\mu; \alpha)} = \frac{S^*_{n,k}(\alpha)}{\phi(\mu; \alpha)} = \frac{1}{\varphi(\mu; \alpha)}$$

is decreasing. Similarly, the minimum partition of λ is shown decreasing.

Next, such a partition as $\varphi(\mu; \alpha) \to 1$ ($\alpha \to 1$) is unique in $\mathcal{P}_{n,k}$. Because of (1.5) or (1.6), $S^*_{n,k}(\alpha)$ is a polynomial in α of degree $n - k$ with single root $\alpha = 1$, and the other real roots are larger than 1. Moreover, $\phi(v; \alpha)$ has the factor $(1 - \alpha)^{\sum_{j>1} s_j}$ (size index expression), and

$$\phi(v; \alpha)/S^*_{n,k}(\alpha) \to 1 \ (\alpha \to 1), \text{ provided that } \sum_{j>1} s_j = 1.$$

The last condition is satisfied only by $\mu = (n - k + 1, 1^{k-1})$.

If the pmf of a partition is decreasing, according to Theorem 1.1 below, it is straightforward that partitions smaller ($\prec$) than it has a decreasing pmf. The existence of a partition with a monotone pmf other than μ or λ and the unimodality of other pmfs are open problems. $\square$

The fact that the mle does not exist at the tails of α is shown asymptotically by Mano (see Fig. 5.3 of [13]). Proposition 1.6 is an elemental justification of his result.

Table 1.3 provides a small numerical example of the likelihood and monotone tail partitions. In the "likelihood" rows, $\searrow$ and $\nearrow$ mean that likelihood is decreasing or increasing, respectively, while the numbers represent the maximum likelihood estimates.

The majorization order of partitions in $\mathcal{P}_{12,5}$ is as follows. The partitions are increasing upward and rightward:

$$
\begin{array}{ccccc}
 & & & & \prec 81^4 \\
 & & & & \prec 721^3 \\
 & & & \prec 62^21^2 \; \prec 631^3 \\
 & & \prec 52^31 & \prec 5321^2 \; \prec 541^3 \\
 & \prec 42^4 \; \prec 432^21 & \prec 43^21^2 \; \prec 4^221^2 \\
3^22^3 & \prec 3^321 \\
\end{array}
$$

The pmf of the partitions of $\mathcal{P}_{16,6}$ is illustrated in Fig. 1.3, and the corresponding sf and df are illustrated in Fig. 1.4. The abscissa of these figures is in a total order Ψ, which will be introduced in Paragraph *Total order and visualization*. The ordinate of Fig. 1.3 is in reverse lexicographical order, $|\mathcal{P}_{16,6}|$ - rlg.

Let $\varphi(s; \alpha) = \varphi(s; n, k; \alpha)$ denote the pmf of GBRP.

Theorem 1.1 *The pmf $\varphi(\nu; \alpha)$, $\nu \in (\mathcal{P}_{n,k}, \prec)$, is totally positive with degree 2, TP2:*

$$
T(\nu; \alpha_0, \alpha_1) := \frac{\varphi(\nu; \alpha_1)}{\varphi(\nu; \alpha_0)}, \quad -\infty < \alpha_0 < \alpha_1 < 1, \tag{1.9}
$$

is increasing in $\nu \in (\mathcal{P}_{n,k}, \prec)$.

Proof It is enough to show that

$$
\frac{\varphi(\mu; \alpha)}{\varphi(\lambda; \alpha)}, \quad \forall \lambda \prec \mu, \; \lambda, \mu, \in (\mathcal{P}_{n,k}, \prec),
$$

Table 1.3 Monotone and unimodal pmf of $\mathcal{P}_{12,5}$, $|\mathcal{P}_{12,5}| = 13$. In the rows of mle, $\searrow$ and $\nearrow$ show that the likelihood is decreasing and increasing, respectively

rlg	1	2	3	4	5	6	7
partition	3^22^3	3^321	42^4	432^21	43^21^2	4^221^2	52^31
mle	$\searrow$	$\searrow$	$\searrow$	$\searrow$	-11.02	-1.004	-3.523

rlg	8	9	10	11	12	13	
partition	5321^2	541^3	62^21^2	631^3	721^3	81^4	
mle	-0.641	0.585	0.100	0.671	0.734	$\nearrow$	

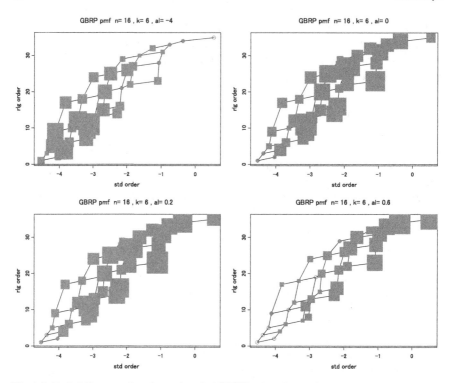

Fig. 1.3 Probability mass functions $\varphi(v; \alpha)$ of GBRP (α) on $\mathcal{P}_{16,6}$, $|\mathcal{P}_{16,6}| = 35$. The area of gray squares is proportional to a pmf. Segments show the adjacency of partitions. The leftmost lowest square represents $3^4 2^2$, and the rightmost highest one does 11.1^5. Upper: $\alpha = -4.0,\ 0.0$; Lower: $\alpha = 0.2,\ 0.6$

is increasing in $\alpha \in (-\infty, 1)$. Such a pair (λ, μ) is specified, by Proposition 1.4, to those that are associated with Moves A1 and A2. In §1.1.3, the change of a partition by Moves A1 and A2 is expressed in the size index: (1.2) and (1.3). That is, the size index $(\tau_1, \ldots, \tau_n)$ changes, by Move A1, to $(\ldots, \tau_{j-1} + 1, \tau_j - 2, \tau_{j+1} + 1, \ldots), 1 < j$.

Hence, for Move A1,

$$\frac{\varphi(\mu; \alpha)}{\varphi(\lambda; \alpha)} = \frac{(1 - \alpha| - 1)_{j-2}(1 - \alpha| - 1)_j}{((1 - \alpha| - 1)_{j-1})^2} = \frac{j - \alpha}{j - 1 - \alpha}, \quad \uparrow (\alpha \uparrow).$$

For Move A2, the result is $(\ldots, \tau_{i-1} + 1, \tau_i - 1, \ldots, \tau_j - 1, \tau_{j+1} + 1, \ldots), 1 < i < j < n - k$, and

$$\frac{\varphi(\mu; \alpha)}{\varphi(\lambda; \alpha)} = \frac{(1 - \alpha| - 1)_{i-2}(1 - \alpha| - 1)_j}{(1 - \alpha| - 1)_{i-1}(1 - \alpha| - 1)_{j-1}} = \frac{j - \alpha}{i - 1 - \alpha}, \quad \uparrow (\alpha \uparrow).$$

$\square$

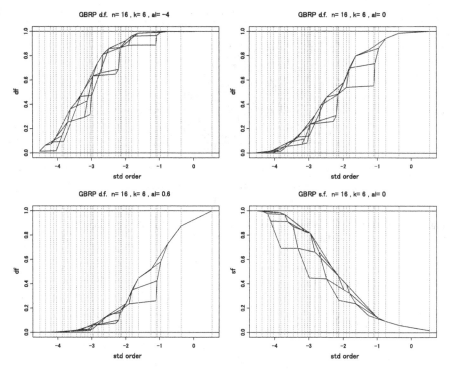

Fig. 1.4 Distribution functions and survival functions of GBRP, $v \in \mathcal{P}_{16,6}$. The top line: df for $\alpha = -4.0, 0.0$; the bottom line: df for $\alpha = 0.6$ and sf for $\alpha = 0.0$. Vertical broken lines show $\Psi(v)$

That is, φ is totally positive of order 2 in the majorization order sense. In other words, φ is Schur-convex in terms of the majorization theory. The related implications will be discussed in the next section.

Total orders and visualization

Theorem 1.1 shows that $T(v; a_0, a_1)$, a function in v depending on arbitrarily chosen $-\infty < a_0 < a_1 < 1$, defines a total (linear) order of $v \in (\mathcal{P}_{n,k}, \prec)$, which is more useful than the reverse lexicographical order for dealing with GBRP. Here, letter a is used instead of α, because this paragraph is apart from inference. By changing (a_0, a_1), the order of incomparable ($\prec$) partitions changes, but its effect is not clear. We propose the simplest one in Proposition 1.7.

A standard selection of (a_0, a_1) is as follows:

Proposition 1.7

$$\Psi(s) := -\sum_{j=1}^{n} s_j \psi(j), \quad \left(\psi(z) = \frac{d}{dz} \log \Gamma(z) \right),$$

is increasing in $v \in (\mathcal{P}_{n,k}, \prec)$.

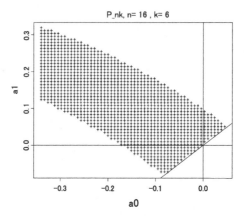

Fig. 1.5 $(a0, a1)$ where the orders by $T(v; a0, a1)$ and by $\Psi(v)$ are the same for $v \in \mathcal{P}_{16,6}$. Lines are of $a0 = 0$, $a1 = 0$, and $a0 = a1$

Proof From Theorem 1.1,

$$\log T(v; a_0, a_1) = \sum_{j=1}^{n} s_j (\log \Gamma(j - a_1) - \log \Gamma(j - a_0)) - k(\log \Gamma(1 - a_1)$$

$$- \log \Gamma(1 - a_0)) - \log S^*_{n,k}(a_1) + \log S^*_{n,k}(a_0)$$

$$= - \left(\sum_{j=1}^{n} s_j \psi(j - a_0) - k\psi(1 - a_0) \right) \epsilon + o(\epsilon), \quad \epsilon = a_1 - a_0 \to 0,$$

is increasing in $s = v \in (\mathcal{P}_{n,k}, \prec)$. $\Psi(s)$ is its special case $a_0 = 0$. $\qquad\square$

Figure 1.5 shows the region of $(a0, a1)$, where the total order by $T(v; a0, a1)$ is the same as that by $\Psi(v)$ for $v \in \mathcal{P}_{16,6}$. The size of the region is smaller for larger n, and the shape in larger $a1$ depends on k.

Hence, $\Psi(v)$ is a standard way of total ordering $(\mathcal{P}_{n,k}, \prec)$. The abscissa of Figs. 1.3 and 1.4 is of $\Psi(v)$. The total ordering of the partitions of $\mathsf{GBRP}(\alpha)$ by $\Psi(v)$ will be denoted by *std*.

Standardization of $\Psi(v)$

The range of $\Psi(v)$ changes by different (n, k), and standardization

$$\Psi^o(v) := (\Psi(v) - E(\Psi(S))/\sqrt{Var(\Psi(S))}, \quad S \sim \mathsf{GBRP}(\alpha),$$

$$E(\Psi(S)) = \sum_{j=1}^{n-k+1} \psi(j)E(S_j), \quad Var(\Psi(S)) = \sum_{i,j=1}^{n-k+1} \psi(i)\psi(j)Var(S)_{i,j},$$

stabilizes the distribution of $\Psi(S)$. To illustrate the effect of standardization, the nine patterns of dos or the part-blocks of different (n, k) are selected, and for each partition, the probability of the upper set versus $\Psi^o(v)$ is plotted in Fig. 1.6.

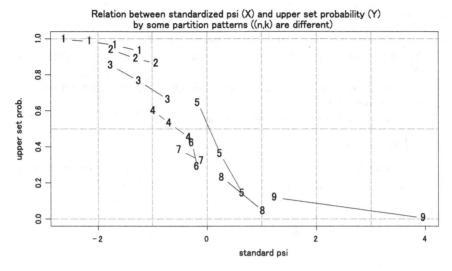

Fig. 1.6 Probability of the upper set versus $\Psi^o(s)$ in several patterns of partitions

The nine patterns, in S/R expression, are as follows:

```
(1) rep(j:1, 1:j), j = 4,5,6,7
```

```
(2) rep(j:1 ,2^(0:(j-1))), j = 4,5,6
```

```
(3) rep(2*(j:1)-1, 1:j), j = 4,5,6
```

```
(4) rep(2*(j:1)^2-1 ,1:j), j =  4,5,6
```

```
(5) rep(j:1, gamma(1:j)), j = 4,5,6
```

```
(6) rep(2^{(j-1):0}, 1:j), j = 4,5
```

```
(7) rep((j:1)^2, 1:j), j = 4,5
```

```
(8) rep(2^{(j-1):0},2^{0:(j-1)}),j=4,5
```

```
(9) rep(gamma(j:1), 1:j), j = 4,5
```

The abscissa of Fig. 1.6 is of $\Psi^o(v)$.

1.3.2 Testing Statistical Hypotheses

One-sided test

One of the motivations of this paper is to test the hypothesis that $\alpha = 0$. An immediate answer is the likelihood ratio test, based on Theorem 1.1.

In the one-parameter family of univariate continuous distributions, a distribution $p(\cdot, \theta)$ belongs to a monotone likelihood ratio family if the ratio $p(x, \theta_1)/p(x, \theta_0)$ is nondecreasing in a function $T(x)$. In Theorem 1.1, the ratio depends on (a_0, a_1), and the uniformly most powerful test does not exist.

In GBRP(α) on $\mathcal{P}_{n,k}$, $1 < k < n$, for testing

$$H : \alpha \le \alpha_0 \text{ versus } K : \alpha > \alpha_0,$$

let $R_v = \mathcal{U}(v)$ be an upper set of a partition v. We observe four facts:

1. R_v is the rejection region of the test with the level $\mathbb{P}\{s \in R_v; \alpha_0\}$.
2.   The power of the test, $\mathbb{P}\{s \in R_v; \alpha\}$, is increasing in α. That is, the test is unbiased.
3.   Randomization test is possible for controlling the level; however, for small n and k, the controllability is limited.
4.   The uniformly most powerful test does not exist.

See [11] for the monotone likelihood family.

There are more than one upper sets that have close levels. For selecting one among them, $\mathcal{U}(v^{(i)})$, $i = 1, 2, \ldots$, a possibility is to select $v^{(i)}$ such as $\Psi(v^{(i)})$ is large. Let $v^{(1)}$ and $v^{(2)}$ be candidates; that is,

$$G(v^{(1)}, \alpha_0) \doteq G(v^{(2)}\alpha_0) \quad \text{and} \quad v^{(1)} \not\prec v^{(2)} \ \& \ v^{(1)} \not\succ v^{(2)}.$$

Define

$$\mathcal{U}^{\cap} := \bigcap_{i=1,2} \mathcal{U}(v^{(i)}), \quad \mathcal{U}^{i\cap} := \mathcal{U}(v^{(i)})\backslash\mathcal{U}^{\cap}, \ i = 1, 2.$$

Any partition in $\mathcal{U}^{1\cap}$ is not comparable with one in $\mathcal{U}^{2\cap}$, and the values of $\Psi(v)$, $v \in \mathcal{U}^{1\cap}$, are smaller than those in $\mathcal{U}^{2\cap}$, as a whole. The probability of these partitions as a function of α is not simple, as observed in the previous subsection. An observation in small examples is that $\mathbb{P}(\mathcal{U}^{i\cap})$, $i = 2$, is larger than that of $i = 1$.

Table 1.4 shows a very simple example of $\mathcal{P}_{16,6}$. The left-hand table shows incomparable two partitions $v^{(i)}$, $i = 1, 2$, with close Ψ values, and the difference sets of their upper set, expressed using

$$\mathcal{U}^{\cap} = \{11.1^5, 10.21^4, 931^4\} \quad \text{or by rlg } \{1, 2, 3\}.$$

In the right-hand table, the probabilities of the set identified by rlg corresponding to those in the left-hand table are compared. The last line represents $\mathbb{P}(\mathcal{U}^{\cap}; \alpha)$.

In Fig. 1.7, the probabilities of three subsets are compared, and the subsets are described in Table 1.5. In this example, the effect of the selection is remarkable.

Table 1.4 Some properties of GBRP (α) on $\mathcal{P}_{16,6}$, $|\mathcal{P}_{16,6}| = 35$. See Figs. 1.3 and 1.4

					$\mathbb{P}(\mathcal{U}(v^{(i)}\backslash\mathcal{U}^{\cap}; \alpha)$		
$v^{(i)}$	rlg	$\Psi(v^{(i)})$	$\mathcal{U}(v^{(i)})\backslash\mathcal{U}^{\cap}$	rlg	$\alpha = 0.$	$\alpha = 0.5$	$\alpha = 0.8$
$92^2 1^3$	4	-1.2546	$\{4\}$	4	0.04797	0.07000	0.04710
$6^2 1^4$	8	-1.1034	$\{5, 8\}$	8	0.05165	0.09773	0.11826
				$\cap$	0.09085	0.27120	0.61039

Table 1.5 Details of Fig. 1.7, $|\mathcal{P}_{40.16}| = 1530$. For a partition v, std ord. = rank($\Psi(v)$)

lines	$v^{(i)}$	rlq	std ord.	$\Psi(v^{(i)})$	$|\mathcal{U}(v^{(i)})\backslash\mathcal{U}^{\cap}|$	$|\mathcal{U}(v^{(i)})|$
chain	$14.43^2 2^4 1^8$	187	935	-2.7780	34	179
broken	$13.63^2 2^3 1^9$	236	1056	-2.1511	55	200
solid	$12.5^2 3^2 21^{10}$	336	1090	-1.9511	95	240

$\mathbb{P}(\mathcal{U}(v^{(i)})\backslash\mathcal{U}^{\cap})$				begn		
rlg	$\alpha = 0.$	$\alpha = 0.5$	$\alpha = 0.8$		$\mathbb{P}(\mathcal{U}(v^{(i)})) = G(v^{(i)})$	
$\mathcal{U}^{\cap}$	0.02671	0.22423	0.71941	rlg	$\alpha = 0.$ $\alpha = 0.5$ $\alpha = 0.8$	
187	0.02393	0.05120	0.01109	187	0.050645 0.27543 0.73050	
236	0.02259	0.07913	0.06998	236	0.049299 0.30337 0.78939	
336	0.02290	0.10445	0.12230	336	0.049613 0.32868 0.84171	

Two-sided test

Now, for testing the hypothesis

$$H : \alpha = \alpha_0 \text{ versus } K : \alpha \neq \alpha_0,$$

let the rejection area be $R = R^+ \bigcup R^-$, where $R^+ = \mathcal{U}(\mu)$, $R^- = \mathcal{L}(\lambda)$, $\lambda \prec \mu$, and $R^+ \bigcap R^- = \emptyset$. The power of the test is

$$\mathbb{P}\{R\} = F(\lambda; \alpha) + G(\mu; \alpha),$$

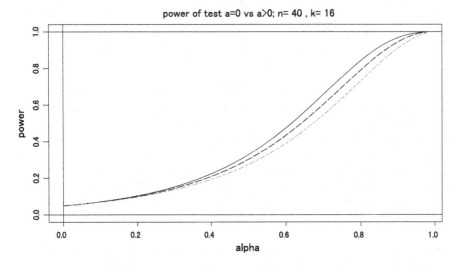

Fig. 1.7 Comparison of rejection regions in $\mathcal{P}_{40.16}$. See Table 1.5 for details

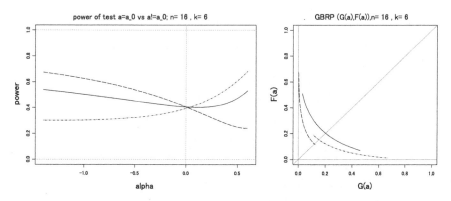

Fig. 1.8 Power of two-sided tests of $\mathcal{P}_{16,6}$. Solid line:(1,16); broken line:(9,29); chain:(8,21)

Table 1.6 Details of Fig. 1.8. $R^- = \mathcal{L}(\lambda)$; $R^+ = \mathcal{U}(\mu)$. $\mathbb{P}(\cdot)$ is evaluated at $\alpha = 0$

Line type	λ	Std ord.	$\mathbb{P}(R^-)$	μ	Std ord.	$\mathbb{P}(R^+)$	$\mathbb{P}(R)$	α_0
Solid line	$3^4 2^2$	1	0.1211	62^5	16	0.1195	0.2406	0.59
Broken line	532^4	9	0.1835	$72^4 1$	29	0.1180	0.3015	-1.30
Chain	$4^2 2^4$	8	0.2313	$632^3 1$	21	0.1714	0.4027	0.11

where F is the distribution function, and G is the survival function.

Proposition 1.8 (conjecture) *If* $F(\lambda; \alpha_1) + G(\mu; \alpha_1) < F(\lambda; \alpha_2) + G(\mu; \alpha_2)$, *for some* (α_1, α_2) *such as* $\alpha_1 < \alpha_2$, *then* $F(\lambda; \alpha_2) + G(\mu; \alpha_2) < F(\lambda; \alpha_3) + G(\mu; \alpha_3)$, *for any* α_3 *such as* $\alpha_2 < \alpha_3 < 1$.
The conjugate statement assumes the inequality for $\alpha_2 < \alpha_1$, *and the induced inequality holds for* $-\infty < \alpha_3 < \alpha_2$.

The proposition cannot be proved by the TP2 property. Provided that the proposition is true for any $\lambda \prec \mu$, such that $F(\lambda; \alpha) + G(\mu; \alpha) < 1$, the power function of the test with the rejection region $R = R^+ \bigcup R^-$, where $R^+ = \mathcal{U}(\mu)$ and $R^- = \mathcal{L}(\lambda)$, is anti-unimodal (a bathtub shape). That is, this is unbiased for testing $H : \alpha = \alpha_0$ versus $K : \alpha \neq \alpha_0$, where $\alpha_0 = \text{argmin}_\alpha (F(\lambda; \alpha) + G(\mu; \alpha))$.

The left part of Fig. 1.8 shows the power of the three arbitrarily selected rejection regions. Their (λ, μ) and the probabilities are described in Table 1.6. The right part of the figure shows the convexity of $(G(\alpha), F(\alpha))$, which supports the proposition.

Tests based on Ψ

Since $\Psi(\nu)$ is increasing in $\nu \in (\mathcal{P}_{n,k}, \prec)$, its behavior would be similar to that of ν. Based on the author's experience, tests based on $\Psi(\nu)$ are more powerful than those

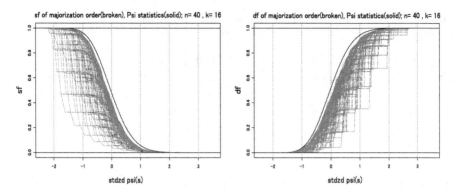

Fig. 1.9 Survival function and distribution function of majorization order (gray) and $\Psi(v)$ test statistics (black) of partitions of $\mathcal{P}_{40,16}$

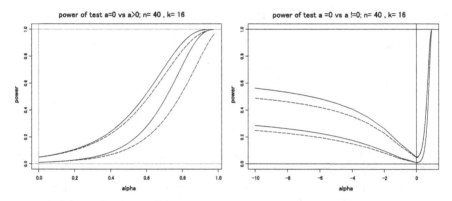

Fig. 1.10 Power of tests based on $\Psi(v)$ (solid lines) and $\mathcal{L}(\lambda) \bigcup \mathcal{U}(\mu)$ (broken lines). Upper two lines are 5% level tests, and lower two lines are 1% tests. Left: one-sided test. Right: two-sided test

based on $\mathcal{U}(v)$ and $\mathcal{L}(v)$, since the latter do not always include partitions having a larger probability at $|\alpha| \gg 0$.

If the level is smaller, the difference between the two types of tests is smaller, since the partitions of the rejection regions of two types are small and similar. When $|\mathcal{P}_{n,k}|$ is large, the rejection regions of two types are also similar, since many partitions with small probabilities are included in both types. See Fig. 1.9.

Example

$\mathcal{P}_{40,16}, |\mathcal{P}_{40,16}| = 1530$, and $H_0 : \alpha = 0$

One-sided test: The power functions are illustrated in Fig. 1.10 on the left.

Two-sided test: The power functions are illustrated in Fig. 1.10 on the right, and the specification of the tests is shown in Table 1.7.

Table 1.7 Details of Fig. 1.10 on the right. $R := R^+ \bigcup R^-$. Comparison of the test based on majorization order, $R^- = \mathcal{L}(\lambda)$ and $R^+ = \mathcal{U}(\mu)$, with the test based on $R^- = \{v; \Psi(v) \leq \Psi(v^{(\ell)})\}$ and $R^+ = \{v; \Psi(v) \geq \Psi(v^{(u)})\}$

Type of test	Actual level	rlg	$\mathbb{P}(R^-)$	rlg	$\mathbb{P}(R^+)$
5% majorization	0.04935	272	0.02269	1225	0.02466
5% Ψ	0.05016	183	0.02514	1289	0.02502
1% majorization	0.009668	141	0.004609	1368	0.005059
1% Ψ	0.010177	120	0.005156	1429	0.005021

Applications

In this paragraph, the probability of the upper set and the standardized deviation of the observed partition of real datasets are calculated, i.e., $\Psi^o(s) = (\Psi(s) - E(\Psi(S)))/S.D.(\Psi(S))$. The parameter of underlying GBRP is assumed to be $\alpha = 0$.

1. Engen [4], p. 106, Christmas counts of bird species in Sweden.

   $\lambda \in \mathcal{P}_{109,34}$

i	1	2	3	4	5	6	7	8	9
j_i	15	12	10	7	5	4	3	2	1
s_{j_i}	1	1	1	2	1	3	5	6	14

   $|\mathcal{P}_{109,34}| = 7902973$, $|\mathcal{U}(\lambda)| = 4018209$, $\mathbb{P}(\mathcal{U}(\lambda)) = 0.298763$, $\Psi(s) = -15.48$, $\Psi^o(s) = -0.1328$.

2. A hoard of Indo-Greek coins, in American numismatic society bibliography (1974), used by Esty WW (1983) Ann Statist 11:905–912, and Holst L (1981) Scand J Statist 8:243–246.

   $\lambda \in \mathcal{P}_{204,178}$

i	1	2	3	4
j_i	4	3	2	1
s_{j_i}	1	2	19	156

   $|\mathcal{P}_{204,178}| = 2436$, $|\mathcal{U}(\lambda)| = 2419$, $\mathbb{P}(\mathcal{U}(\lambda)) = 0.401952$, $\Psi(s) = 78.911$, $\Psi^o(s) = -0.00590$.

3. Baayen [1], p. 310, Plural nouns in *The Bloody Wood* by M. Innes, from TOSCA corpus by Keulen, 1986.

Table 1.8 $S^*_{n,1}(\alpha) = (1 - \alpha| - 1)_{n-1};\quad \alpha = 0.2.$

n	165	166	167	168	169	170	171	172
$S^*_{n,1}$	$1.018 \cdot 10^{293}$	$1.677 \cdot 10^{295}$	$2.781 \cdot 10^{297}$	$4.638 \cdot 10^{299}$	$7.783 \cdot 10^{301}$	$1.314 \cdot 10^{304}$	$2.231 \cdot 10^{306}$	Inf

$\lambda \in \mathcal{P}_{386,65}$

i	1	2 3 4 5 6 7	8	9	10
j_i	17	11 9 7 6 5 4	3	2	1
s_{j_i}	1	1 1 1 3 5 4	16	31	173

$|\mathcal{P}_{386,65}| = 35517905970066664,\ E(\Psi(S)) = 27.99732,\ Var(\Psi(S)) = 15.16550,\ \Psi(s) = 45.145343,\ \Psi^o(s) = 4.403375.$ (The calculation of the upper set failed due to its size.)

1.3.3 Problems in Computation

To calculate $\varphi(v; \alpha)$ for all $v \in \mathcal{P}_{n,k}$, the computation of the normalization factor $S^*_{n,k}(\alpha)$ is not necessary. That is just the sum of terms without the factor. For computing the pmf of partitions of the larger part of $v \in \mathcal{P}_{n,k}$, $S^*_{n,k}(\alpha)$ is necessary. However, the value of $S^*_{n,k}(\alpha)$ overflows easily when n and k increase, as shown in Table 1.8. Note that $\Gamma(172) = $ Inf and $10^{309} = $ Inf.

Computation of $S^*_{n,k}(\alpha)$

One way to avoid the overflow is to use the pmf (1.5) of **EPSF-K**:

$$f_n(k; \theta, \alpha) = \frac{(\theta| - \alpha)_k}{(\theta| - 1)_n} S^*_{n,k}(\alpha)$$

$$= \frac{\Gamma(\theta)}{\Gamma(\theta + n)} S^*_{n,k}(\alpha) \times \begin{cases} \frac{\Gamma(\theta/\alpha+k)\alpha^k}{\Gamma(\theta/\alpha)}, & \text{if } \alpha > 0, \\ \theta^k, & \text{if } \alpha = 0, \\ \frac{\Gamma(\theta/(-\alpha)+1)(-\alpha)^k}{\Gamma(\theta/(-\alpha)-(k-1))}, & \text{if } \alpha < 0, \theta = -m\alpha, m = 2, 3, \ldots. \end{cases}$$

Note that

$$S^*_{n,k}(0) = \begin{bmatrix} n \\ k \end{bmatrix}, \quad \text{the unsigned Stirling numbers of the first kind.}$$

Table 1.9 Numerical values of pmf $f_n(k; 1, 0)$ at the right tail

$n = 180$	$k = 176$	$k = 177$	$k = 178$	$k = 179$
$f_n(k) =$	1.336e-314	3.392e-318	6.423e-322	0.0e+0
$n = 200$	$k = 183$	$k = 184$	$k = 185$	$k = 186$
$f_n(k) =$	1.683e-317	1.608e-320	9.881e-324	0.0e+0
$n = 250$	$k = 192$	$k = 193$	$k = 194$	$k = 195$
$f_n(k) =$	9.279e-318	2.385e-320	5.435e-323	0.0e+0

The pmf $f_n(k; \theta, \alpha)$ is calculated by recurrence. The asymptotics of $\begin{bmatrix} n \\ k \end{bmatrix}$ is studied by Wilf [22], Hwang [10], Louchard [12], and Grünberg [9]. The computation of $S_{n,k}^*(\alpha)$ by the holonomic gradient methods was developed by Mano [13].

The value of θ is arbitrary, and $\theta = 1$ is preferable due to simplicity. However, to avoid the underflow as seen in Table 1.9, one possible way is to let k be the mode of EPSF-K (θ, α).

Proposition 1.9 *The value of θ, for which the mode of* EPSF-K (θ, α) *is k, is the solution of*

$$\psi(\theta + n) - \psi(\theta) - \frac{1}{\alpha}\left(\psi\left(\frac{\theta}{\alpha} + k\right) - \psi\left(\frac{\theta}{\alpha}\right)\right) = 0, \quad if \, \alpha > 0,$$

$$\theta(\psi(\theta + n) - \psi(\theta + 1)) - k = 0, \quad if \, \alpha = 0,$$

where $\psi(\theta)$ is the digamma (or psi) function.

The proposition is derived by setting the derivative of (1.5) to 0.

Example

The mode of $f_n(k; \theta, 0)$:

$n = 180$	$k = 170$	$k = 175$	$n = 200$	$k = 185$	$k = 190$	$n = 250$	$k = 190$	$k = 195$
$\theta =$	1671.35	3988.28	$\theta =$	1290.01	2079.12	$\theta =$	368.215	416.43

Acknowledgements The author expresses his gratitude to the referee for his efforts to check the details of the paper. Because of his comments, the paper is much improved and made readable.

Table 1.10 Classification of *downward* transform types from a partition $\{(j_i, s_{j_i}), 1 \le i \le b\}$. The type of part-blocks, from which one element is to be taken out, is classified. C: within $\mathcal{P}_{n,k}$, D: from $\mathcal{P}_{n,k}$ to $\mathcal{P}_{n,k+1}$

j_i	C1,D1: $j_i - j_{i+1} > 1$	C2,D2: $j_i - j_{i+1} = 1$
C	$i \le b-1, b \ge 2$	$i \le b-2, b \ge 3;\ j_{i+1} - j_{i+2} = 1$
D	$i = b$	$i = b-1$

Change of part-blocks in each transform	
C1 $j_i - j_{i+1} > 1$	C2 $j_i - j_{i+1} = 1$ (and $j_{i+1} - j_{i+2} = 1$)
$j_i - j_{i+1} > 2$	$j_i - j_{i+1} = j_{i+1} - j_{i+2} = 1$
$\cdots (j_i, s_{j_i} - 1)(j_i - 1, 1)(j_{i+1} + 1, 1)(j_{i+1}, s_{j_{i+1}} - 1) \cdots$	$\cdots (j_i, s_{j_i} - 1)(j_{i+1}, s_{j_{i+1}} + 2)(j_{i+2}, s_{j_{i+2}} - 1) \cdots$
$j_i - j_{i+1} = 2$	$j_i - j_{i+1} = j_{i+1} - j_{i+2} = 1$
$\cdots (j_{i+1} + 1, 1)(j_{i+1}, s_{j_{i+1}} - 1) \cdots$	$\cdots (j_{i+1}, s_{j_{i+1}} + 2)(j_{i+2}, s_{j_{i+2}} - 1) \cdots$

D1: $i = b, s_b > 2$	D2: $i = b-1, j_i = 2, s_{j_b} = s_1 > 1$
$j_b > 2,\ s_{j_b} > 1$	$j_b = 1,\ s_{j_b} > 0$
$\cdots (j_b, s_{j_b} - 1)(j_b - 1, 1)(1, 1).$	$\cdots (2, s_2 - 1)(1, s_1 + 2).$
$j_b \ge 2,\ s_{j_b} = 1$	$j_b = 1,\ s_{j_{b-1}} = 1$
$\cdots (j_b - 1, 1)(1, 1).$	$\cdots (1, s_1 + 2).$

Appendix

Table of lower adjacency

In §1.1.3, upward adjacency is discussed, and downward adjacency is shown in Table 1.10. Comparing it with Table 1.2, we note that conceptually upward and downward adjacency are reverse, but not the corresponding moves.

Majorization order of integer vectors

A partition $c = \{c_1, \ldots, c_k\}, c \vdash n$, with part-blocks $(j_i, s_{j_i}), 1 \le i \le b$, corresponds to $\binom{k}{s_{j_1}, \ldots, s_{j_b}}$ positive integer vectors $\mathbf{c} = (c_1, \ldots, c_k) \in \mathbb{N}^k$ in such a way that, if the axes are neglected and the components are mixed, the vectors reduce to $(c_1, \ldots, c_k) \downarrow$. Actually, the set of k-vectors $\mathbf{c}, c \vdash n$, is a simplex:

$$\Delta_k^+(n) := \left\{ \mathbf{c}; \mathbf{c} \in \mathbb{N}^k, \sum_{j=1}^{k} c_j = n \right\}, 1 \le k \le n.$$

Similarly, the size index expression $s = (s_1, \ldots, s_n), s \in \mathcal{P}_n$, corresponds to $\binom{n}{n-k, s_1, \ldots, s_b} = \binom{n}{k} \binom{k}{s_1, \ldots, s_b}$ nonnegative integer vectors $\mathbf{s} = (s_1, \ldots, s_n) \in \mathbb{N}_0^n$. The set of n-vectors $\mathbf{s}, s \in \mathcal{P}_n$, is a simplex:

$$\Delta^0(n) := \left\{ \mathbf{s}; \mathbf{s} \in \mathbb{N}_0^n, \sum_{j=1}^{n} j s_j = n \right\}.$$

Note that $\Delta_k^+(n) \subset \Delta^0(n)$, $1 \leq k \leq n$, and $\Delta^0(n) \subset \{\mathbf{z} \in \mathbb{Z}^n, z_1 + \cdots + z_n = n\}$. More equal partitions are in the inner part, and less equal ones are in the outer part of $\Delta_k^+(n)$ and $\Delta^0(n)$. The standard theory of majorization deals with ordering of $\mathbb{R}^n$, and there are several equivalent definitions and some extensions.

To find adjacent partitions, we could consider a dual way: (C) Move to a lower partition of $\mathcal{P}_{n,k}$ and (D) Move from the partitions of $\mathcal{P}_{n,k}$ to the partitions of $\mathcal{P}_{n,k+1}$. The moves from inequality to equality were studied by Lorenz, Pigou, Dalton, and others in the beginning of the twentieth century, and called *the Principle of Transfers*. See Table 1.10 for the summary of the moves.

A real-valued function $\phi(x)$, $x \in \mathbb{R}^n$, such that $\phi(x) \leq \phi(y)$ if $x \prec y$, is called *Schur-convex*, which actually means increasing. Various inequalities are systematically obtained from Schur-convex functions, and the characterization of Schur-convex functions is the principal part of the majorization theory.

Partial quantile

As mentioned in introduction, the purpose of this paragraph is to investigate shortly the notions of the partial quantile theory in GBRP.

Let X be a random variable on a poset $(\mathcal{X}, \leq)$. For any element $x \in \mathcal{X}$, in terms of its lower and upper set,

$$\mathcal{L}(x) := \{w \in \mathcal{X}; w \leq x\} \quad \text{and} \quad \mathcal{U}(x) := \{w \in \mathcal{X}; w \geq x\},$$

define *the set of elements comparable* with x as

$$C(x) := \mathcal{L}(x) \cup \mathcal{U}(x),$$

and *the probability of comparison* as

$$p_x := \mathbb{P}\{X \in C(x)\}.$$

The basic idea of partial quantile is to deal with the distribution functions and the survival functions conditional to $C(x)$;

$$\mathcal{F}(x) := \mathbb{P}\{X \in \mathcal{L}(x)|\ X \in C(x)\} \quad \text{and} \quad \mathcal{S}(x) := \mathbb{P}\{X \in \mathcal{U}(x)|\ X \in C(x)\}.$$

For a fixed element $x \in \mathcal{X}$, the interval $(l(x), u(x)) \subset (0, 1)$ is defined by

$$(l(x), u(x)) := (1 - \mathcal{S}(x), \mathcal{F}(x)) = \{\tau \in (0, 1); \mathcal{F}(x) > \tau\ \&\ \mathcal{S}(x) > 1 - \tau\}.$$

Conversely, the set of elements such as its interval covers $\tau \in (0, 1)$,

$$Q(\tau) := \{x \in \mathcal{X}; (l(x), u(x)) \ni \tau\} = \{x \in \mathcal{X}; 1 - \mathcal{S}(x) < \tau < \mathcal{F}(x)\},$$

is *the τ-partial quantile surface*.

For specifying a quantile among those on the surface, the maximizer of the probability of comparison is selected as

$$x_\tau := \mathrm{argmax}_{x \in Q(\tau)} \; p_x.$$

This is the [2] definition of *the τ-partial quantile.*

Figure 1.11 illustrates the intervals $(l(x), u(x))$ of GBRP ($\alpha = 0$) on $x \in \mathcal{P}_{12,5}$ with the majorization order. A vertical line at $\tau \in (0, 1)$ crosses some horizontal segments $(l(x), u(x))$, and the set of crossing x's is the τ-partial quantile interface $Q(\tau)$ (the naming is meaningless in the present context). Table 1.11 provides the

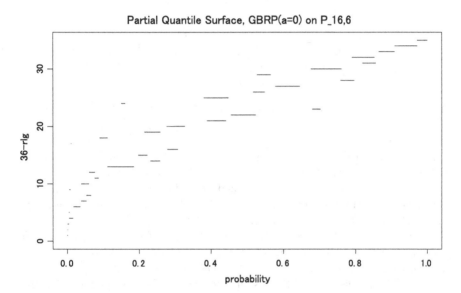

Fig. 1.11 Partial quantile surface; $n = 16$ and $k = 6$

Table 1.11 Details of a part of Fig. 1.11

partition	rlg order	$l(x)$	$u(x)$	p_x
$732^2 1^2$	11	0.4747	0.5483	0.8379
6521^3	14	0.5507	0.6164	0.8762
6431^3	15	0.4810	0.5340	0.9036
$642^2 1^2$	16	0.3546	0.4115	0.9477

rlg order	(l, u)
16	0.3546 0.4115
	0.4115 0.4747
11	0.4747 0.4810
15	0.4810 0.5340
11	0.5340 0.5483
	0.5483 0.5507
14	0.5507 0.6164

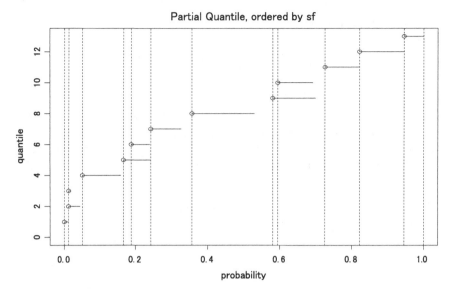

Fig. 1.12 Partial quantile ordered by sf; $n = 12$ and $k = 5$

details of a part of Fig. 1.11 showing the interval $(l(x), u(x))$ and the comparison probability p_x of four partitions with rlg sequence numbers 11, 14, 15, and 16.

Note that between the intervals of 16 and 11, and between those of 11 and 14, there are gaps, and for $\tau \in (0.4115, 0.4747) \cup (0.5483, 0.5507)$, no quantile is determined. There are other similar gaps seen in Fig. 1.11.

The interval of 15 lies completely inside within that of 11, and the comparison probability of 15 is larger than that of 11. Hence the partition of 11 is selected as a partial quantile for separate intervals, $\tau \in (0.4747, 0.4810) \cup (0.5340, 0.5483)$.

A possible approach for GBRP is simply to use $\tau = \mathcal{F}(x)$ or $\mathcal{S}(x)$ to define the quantile, thus giving up to reconcile the difference in using them. Figure 1.12 shows the quantile, based on $\mathcal{S}(x)$, of GBRP(α) on $\mathcal{P}_{12,5}$ for $\alpha = 0$. The partitions $x \in \mathcal{P}_{12,5}$ are totally ordered by the value of $1 - \mathcal{S}(x)$, shown by circlets in Fig. 1.12. For the values of τ in an interval between the adjacent vertical lines, the partition whose circlet is on the left vertical line is the τ-partial quantile.

References

1. Baayen RH (2001) Word frequency distributions. Kluwer, Dordrecht
2. Belloni A, Winkler RL (2011) On multivariate quantiles under partial orders. Ann Statist 39:1125–1179. https://doi.org/10.1214/10-AOS863
3. Crane H (2016) The ubiquitous Ewens sampling formula. Statist Sci 31(1):1–19, with discussions and rejoinder

4. Engen S (1978) Stochastic abundance models, with emphasis on biological communities and species diversity. Chapman and Hall, London
5. Feng S (2010) The Poisson-Dirichlet distribution and related topics. Springer, Berlin
6. Ferguson TS (1973) A Bayesian analysis of some nonparametric problems. Ann Statist 1:209–230
7. Ghosh JK, Ramamoorthi RV (2003) Bayesian nonparametrics. Springer, New York
8. Gnedin A, Pitman J (2006) Exchangeable Gibbs partitions and Stirling triangles. J Math Sci 138:5699–5710. (Translation from Russian paper 2005)
9. Grünberg DB (2006) Asymptotic expansions for the Stirling numbers of the first kind. Results Math 49:89–125. https://doi.org/10.1007/s00025-006-0211-7
10. Hwang HK (1995) Asymptotic expansions for the Stirling numbers of the first kind. J Comb Theo Ser A 71(2):343–351
11. Lehmann EL, Romano JP (2005) Testing statistical hypotheses. Springer, New York
12. Louchard G (2010) Asymptotics of the Stirling numbers of the first kind revisited. Discr Math Theo Compu Sci 12(2):167–184
13. Mano S (2018) Partitions, hypergeometric systems, and Dirichlet processes in statistics. Springer, Tokyo. https://doi.org/10.1007/978-4-431-55888-0
14. Marshall AW, Olkin I, Arnold BC (2011) Inequalities: Theory of majorization and its applications, 2nd edn. Springer, New York
15. Pitman J (2006) Combinatorial stochastic processes. Lec Notes Math 1875, Springer, Berlin
16. Rosenbaum P (1991) Some poset statistics. Ann Statist 19:1091–1097
17. Shaked M, Shanthikumar JG (2007) Stochastic orders. Springer, New York
18. Sibuya M (1993) A random clustering process. Ann Inst Statist Math 45:459–465
19. Sibuya M (2014) Prediction in Ewens-Pitman sampling formula and random samples from number partitions. Ann Inst Statist Math 66:833–864. https://doi.org/10.1007/s10463-013-0427-8
20. Takayama N, Kuriki S, Takemura A (2018) A-hypergeometric distributions and Newton polytopes. Adv Appl Math 99:109–133
21. Tavaré S, Ewens WJ (1997) Multivariate Ewens distribution, Chap 41. In: Kotz J, Balakrishnan N (eds) Discrete multivariate distributions. Wiley, pp 232–246
22. Wilf HS (1993) The asymptotic behavior of the Stirling numbers of the first kind. J Comp Theo Series A 64:344–349

Chapter 2
Asymptotic and Approximate Discrete Distributions for the Length of the Ewens Sampling Formula

Hajime Yamato

Abstract The Ewens sampling formula is well known as the probability for a partition of a positive integer. Here, we discuss the asymptotic and approximate discrete distributions of the length of the formula. We give a sufficient condition for the length to converge in distribution to the shifted Poisson distribution. This condition is proved using two methods: One is based on the sum of independent Bernoulli random variables, and the other is based on an expression of the length that is not the sum of independent random variables. As discrete approximations of the length, we give those based on the Poisson distribution and the binomial distribution. The results show that the first two moments of the approximation based on the binomial distribution are almost equal to those of the length. Two applications of this approximation are given.

Keywords Approximation · Binomial distribution · Charier polynomial · Chinese restaurant process · Convergence in distribution · Ewens sampling formula · Krawtchouk polynomial · Length of partition · Poisson distribution · Shifted distribution

2.1 Introduction

As the probability of a random unordered partition $(C_1, C_2, \ldots, C_n)$ of an integer n, we consider the Ewens sampling formula (ESF) with the parameter θ, ESF(θ), as follows:

$$P\big((C_1, C_2, \ldots, C_n) = (c_1, c_2, \ldots, c_n)\big) = \frac{\theta^k}{\theta^{[n]}} \cdot \frac{n!}{\prod_{j=1}^k j^{c_j} c_j!} \quad (0 < \theta < \infty),$$

H. Yamato (✉)
Emeritus of Kagoshima University, Take 3-32-1-708, Kagoshima 890-0045, Japan
e-mail: yamato_march@hiz.bbiq.jp

© The Author(s), under exclusive license to Springer Nature Singapore Pte Ltd. 2020
N. Hoshino et al. (eds.), *Pioneering Works on Distribution Theory*,
JSS Research Series in Statistics,
https://doi.org/10.1007/978-981-15-9663-6_2

37

where $\theta^{[n]} = \theta(\theta + 1) \cdots (\theta + n - 1)$, and $c_1, c_2, \ldots, c_n$ are nonnegative integers satisfying $\sum_{i=1}^{n} i c_i = n$ and $\sum_{i=1}^{n} c_i = k$. Ewens [12] discovered the ESF in the context of genetics. Antoniak [2] discovered it independently in a nonparametric Bayes context using Ferguson [14] Dirichlet process. The formula appears in many statistical contexts, including Bayesian statistics, patterns of communication, and genetics (e.g., see [17, Chap. 41]) and [9]). The probability function (p.f.) of the length $K_n = \sum_{i=1}^{n} C_i$ is

$$P(K_n = k) = | s(n, k) | \frac{\theta^k}{\theta^{[n]}} \quad (k = 1, 2, \ldots, n) , \tag{2.1}$$

where $| s(n, k) |$ is the signless Stirling number of the first kind [12].

The relation between $\mathcal{L}(K_n)$ and the Poisson distribution $P_n = Po\big(E(K_n)\big)$ is given by Arratia and Tavaré [3, p. 326]) using the total variation distance, given by $d_{TV}\big(\mathcal{L}(K_n), \mathcal{L}(P_n)\big) \asymp 1/\log n$. Additional relations are given by Arratia et al. [5, 6] for logarithmic combinatorial structures, which include the ESF. Yamato [23] gives the condition for K_n to converge to the shifted Poisson distribution, as well as an approximation for $\mathcal{L}(K_n)$ using the Charier polynomial. Yamato [24, 25] give an approximation for $\mathcal{L}(K_n)$ using the Krawtchouk polynomial. They also give an approximation in which the mean is equal to $E(K_n)$ and the variance is almost equal to $Var(K_n)$.

We first explain K_n and its distribution using the well-known Pitman's Chinese restaurant process (e.g., see [1, (11.19)]) for two forms, (I) and (II):

Chinese restaurant process (I): Consider customers $1, 2, \ldots, n$ arriving sequentially at an initially empty restaurant that has many tables.
(I) The first customer sits at the first table.
(II) The jth customer selects a table as follows ($j = 2, 3, \ldots, n$): the customer selects

$$\text{(1) a new table with probability } \frac{\theta}{\theta + j - 1} ,$$

$$\text{(2) an occupied table with probability } \frac{j - 1}{\theta + j - 1} .$$

Define random variables ξ_j ($j = 1, \ldots, n$) as follows:
$\quad \xi_j := 1$ if the jth customer sits at a new table,
$\quad \xi_j := 0$ if the jth customer sits at an occupied table.
Then, we have

$$P(\xi_j = 1) = \frac{\theta}{\theta + j - 1} , \quad P(\xi_j = 0) = \frac{j - 1}{\theta + j - 1} \quad (j = 1, \ldots, n) . \tag{2.2}$$

After n customers are seated, the number of occupied tables is given by

$$K_n = \xi_1 + \xi_2 + \cdots + \xi_n , \tag{2.3}$$

with a p.f. given by (2.1).

Here, we consider the ordered tables. Let $A_1, A_2, \ldots$ denote the number of customers occupying the first table, second table, and so on. It is easily seen that

$$P(A_1 = a_1, A_2 = a_2, \ldots, A_k = a_k) = \frac{\theta^k}{\theta^{[n]}} \cdot \frac{n!}{a_k(a_k + a_{k-1}) \cdots (a_k + a_{k-1} + \cdots + a_1)},$$
(2.4)

for the positive integers $a_1, a_2, \ldots, a_k$ satisfying $a_1 + a_2 + \cdots + a_k = n$ (see (2.29) in the Appendix).

This distribution can also be derived using a similar model (e.g., see [17]; Sect. 2 of Chap. 41). Let C_1 be the number of tables occupied by one customer, C_2 be the number of tables occupied by two customers, and so on. Then, by (2.33) in the Appendix, $(C_1, \ldots, C_n)$, corresponding to $(A_1, \ldots, A_k)$, has ESF(θ).

Chinese restaurant process (II): The $0-1$ sequence $\xi_1 \xi_2 \cdots \xi_n 1$ ($\xi_1 = 1$) indicates that each customer $(1, 2, \ldots, n)$ sits at a new table (1) or at an occupied table (0). Let

$$C_j := \text{no. of } j - \text{spacings in } \xi_1 \xi_2 \cdots \xi_n 1 \quad (j = 1, \ldots, n) ,$$

which can be written as

$$C_1 = \sum_{i=1}^{n-1} \xi_i \xi_{i+1} + \xi_n , \quad C_n = \xi_1 \bar{\xi}_2 \cdots \bar{\xi}_n ,$$
(2.5)

$$C_j = \sum_{i=1}^{n-j} \xi_i \bar{\xi}_{i+1} \cdots \bar{\xi}_{i+j-1} \xi_{i+j} + \xi_{n-j+1} \bar{\xi}_{n-j+2} \cdots \bar{\xi}_n \quad (j = 2, \ldots, n - 1) ,$$
(2.6)

where $\bar{\xi}_i = 1 - \xi_i$ $(i = 1, 2, \ldots, n)$. Then,

$$K_n = C_1 + C_2 + \cdots + C_n$$
(2.7)

(see [4]; p. 523, ↑ 10).

Here, we consider the ordered spacings. Let $D_1, D_2, \ldots$ be the length of the first spacing, second spacing, and so on. It is easily seen that

$$P(D_1 = d_1, D_2 = d_2, \ldots, D_k = d_k) = \frac{\theta^k}{\theta^{[n]}} \cdot \frac{n!}{d_1(d_1 + d_2) \cdots (d_1 + d_2 + \cdots + d_k)},$$
(2.8)

for the positive integers $d_1, d_2, \ldots, d_k$ satisfying $d_1 + d_2 + \cdots + d_k = n$ (see (2.32) in the Appendix). Then, by (2.33) in the Appendix, $(C_1, C_2, \ldots, C_n)$, corresponding to $(D_1, \ldots, D_k)$, has ESF (θ). This correspondence is also stated in [10, p. 10], where their $D_1, D_2, \ldots$ denote waiting times of the next mutation in their coalescent process. For the property that $(C_1, C_2, \ldots, C_n)$, given by (2.5) and (2.6), has ESF(θ), see, for example, [3, Theorem 1] and [7]; 2nd paragraph. The distribution (2.8) is also given by Ethier [11, Theorem 4.1.].

Ewens [13] refers to the distribution with form (2.4) as the Donnelly–Tavaré–Griffiths formula, quoting his relation (163), which is equivalent to (2.4). However, the distribution with form (2.8) is stated in the cited reference, [10, (4.2)]. Therefore, [22] refers to the distribution with form (2.8) as the Donnelly–Tavaré–Griffiths formula II.

This study aims to explain the asymptotic and approximate discrete distributions for $\mathcal{L}(K_n)$, following [23–25], and to give a new proof of the sufficient condition for K_n to converge to the shifted Poisson distribution.

Using relation (2.3), [23] gives the condition for K_n to converge to the shifted Poisson distribution, which we explain in Sect. 2.2. This convergence is also proved using relation (2.7) in Sect. 2.3. In Sect. 2.4.1, we compare the Poisson approximations to $\mathcal{L}(K_n)$ by Arratia et al. [5, 6] and Yamato [23]. In Sect. 2.4.2, we give the shifted binomial approximations to $\mathcal{L}(K_n)$, following [24, 25]. As an approximation to $\mathcal{L}(K_n)$, we recommend using the shifted binomial approximation, based on its first two moments. Two applications are given in Sect. 2.5. In the Appendix, we explain some of the relations used in the Chinese restaurant processes (I) and (II).

2.2 Convergence in Distribution of K_n (I)

Using Chinese restaurant process (I), we consider K_n, expressed as (2.3), and the independent Bernoulli random variables $\xi_1, \xi_2, \ldots, \xi_n$, with probabilities given by (2.2). Because $\xi_1 = 1$ almost surely (a.s.), we have

$$K_n = 1 + L_n \text{ a.s. where } L_n := \xi_2 + \cdots + \xi_n . \tag{2.9}$$

With regard to the convergence in distribution to the Poisson distribution for the sum of independent Bernoulli random variables, [21, Theorem 3]; gives a sufficient condition using the point-wise convergence of the p.f. This has also been proved using the method of moments and the total variation distance ([23]; Proposition 3.1 and Remark 2.1, 2). This result can be used to obtain L_n, because it is a sum of independent Bernoulli random variables. Noting that $\sum_{j=2}^{n} \theta/(\theta + j - 1) \approx \theta \log n$, we have the following proposition:

Proposition 2.1 (Yamato [23, Proposition 3.1]) *For the number K_n of distinct components of $ESF(\theta)$, it holds that*

$$If \ \theta \log n \to \lambda \ (n \to \infty \ and \ \theta \to 0), \ then \ K_n \xrightarrow{d} 1 + Po(\lambda) \ (0 < \lambda < \infty) , \tag{2.10}$$

where $\xrightarrow{d}$ indicates convergence in distribution.

The convergence in (2.10) is in contrast to the well-known Poisson "law of small numbers" for the sum of independent and identically distributed (i.i.d.) Bernoulli random variables (e.g., see [8, p. 1]). The distribution $1 + Po(\lambda)$ is shifted to the right by one for the Poisson distribution $Po(\lambda)$ and is also known as the size-biased version of $Po(\lambda)$ (e.g., see [6, p. 78, Lemma 4.1]).

The above proof is based on the sum of independent random variables. We prove (2.10) using a different method with Chinese restaurant process (II).

2.3 Convergence in Distribution of K_n (II)

Using Chinese restaurant process (II), we consider K_n, given by (2.7) with (2.5) and (2.6). Eliminating the last term from the right-hand sides of (2.5) and (2.6), for $1, 2, \ldots, n - 1$, we have

$$B_1 = \sum_{i=1}^{n-1} \xi_i \xi_{i+1}, \quad B_j = \sum_{i=1}^{n-j} \xi_i \bar{\xi}_{i+1} \cdots \bar{\xi}_{i+j-1} \xi_{i+j} \quad (j = 2, 3, \ldots, n-1) \,. \quad (2.11)$$

Lemma 2.1 *We can write K_n as follows:*

$$K_n = \sum_{j=1}^{n-1} B_j + R_n \,, \quad (2.12)$$

where

$$R_n = 1 - \prod_{j=1}^{n} \bar{\xi}_j. \quad (2.13)$$

Proof From (2.7) with (2.5), (2.6), and (2.11), we have

$$R_n = 1 - \bar{\xi}_n + \sum_{j=2}^{n} \xi_{n-j+1} \bar{\xi}_{n-j+2} \cdots \bar{\xi}_n \,. \quad (2.14)$$

For the first four terms of the right-hand side of (2.14), we have $1 - \bar{\xi}_n + \xi_{n-1}\bar{\xi}_n + \xi_{n-2}\bar{\xi}_{n-1}\bar{\xi}_n = 1 - \bar{\xi}_{n-1}\bar{\xi}_n + \xi_{n-2}\bar{\xi}_{n-1}\bar{\xi}_n = 1 - \bar{\xi}_{n-2}\bar{\xi}_{n-1}\bar{\xi}_n$. Repeating this methods, we obtain (2.13). □

Because $\bar{\xi}_1 = 0$ a.s., we have $R_n = 1$ a.s., by (2.13). Therefore, from (2.12), we have the following:

Lemma 2.2

$$K_n = \sum_{j=1}^{n-1} B_j + 1 \quad \text{a.s.} \tag{2.15}$$

Here, we put

$$B_j(\infty) = \sum_{i=1}^{\infty} \xi_i \bar{\xi}_{i+1} \cdots \bar{\xi}_{i+j-1} \xi_{i+j} \quad (j = 1, 2, \ldots) . \tag{2.16}$$

Lemma 2.3 (Arratia et al. [4]) *The random variables $B_j(\infty)$ ($j = 1, 2, \ldots$) are independent and follow the Poisson distribution $Po(\theta/j)$.*

The total variation distance d_{TV} between the probability measures Q_1 and Q_2 over $\{0, 1, 2, \cdots\}$ is denoted as follows:

$$d_{TV}(Q_1, Q_2) = \frac{1}{2} \sum_{j=0}^{\infty} |Q_1(j) - Q_2(j)| .$$

Proof of Proposition 2.1 The difference between $B_j(\infty)$ and B_j is

$$B_j(\infty) - B_j = \sum_{i>n-j}^{\infty} \xi_i \bar{\xi}_{i+1} \cdots \bar{\xi}_{i+j-1} \xi_{i+j} \quad (j = 1, 2, \ldots, n-1) .$$

Taking the expectation of the absolute value of the above, we have

$$E \mid B_j(\infty) - B_j \mid < \sum_{i>n-j}^{\infty} E(\xi_i \xi_{i+j})$$

$$= \frac{\theta^2}{j} \sum_{i>n-j}^{\infty} \left(\frac{1}{\theta+i-1} - \frac{1}{\theta+i+j-1} \right) < \frac{\theta^2}{\theta+n-j} .$$

The sums of the above left and right sides over $1 \leq j \leq n-1$ give

$$\sum_{j=1}^{n-1} E \mid B_j(\infty) - B_j \mid < \theta^2 \sum_{j=1}^{n-1} \frac{1}{\theta+j} < \theta^2 \sum_{j=1}^{n-1} \frac{1}{j} . \tag{2.17}$$

For any $\varepsilon > 0$, by Markov's inequality, we have

$$P\left(\left| \sum_{j=1}^{n-1} B_j - \sum_{j=1}^{n-1} B_j(\infty) \right| \geq \varepsilon \right) \leq \frac{1}{\varepsilon} \sum_{j=1}^{n-1} E \left| B_j(\infty) - B_j \right| . \tag{2.18}$$

By (2.17) and (2.18),

$$P\left(\left|\sum_{j=1}^{n-1} B_j - \sum_{j=1}^{n-1} B_j(\infty)\right| \geq \varepsilon\right) < \frac{\theta^2}{\varepsilon} \sum_{j=1}^{n-1} \frac{1}{j} \quad \text{for } \varepsilon > 0 . \tag{2.19}$$

Because $\sum_{j=1}^{n-1} B_j(\infty) \stackrel{d}{=} Po\big(\theta \sum_{j=1}^{n-1}(1/j)\big)$, by Lemma 2.3,

$$d_{TV}\left(\mathcal{L}\left(\sum_{j=1}^{n-1} B_j(\infty)\right), Po(\lambda)\right) \leq \left|\theta \sum_{j=1}^{n-1} \frac{1}{j} - \lambda\right|, \tag{2.20}$$

where, for the inequality, see [15, (8) Lemma]. Here, we note the convergence $\sum_{j=1}^{n-1}(1/j)/\log n \to 1$ as $n \to \infty$. If $\theta \log n \to \lambda$ (as $n \to \infty$ and $\theta \to 0$), then by (2.19) and (2.20), we have

$$\sum_{j=1}^{n-1} B_j - \sum_{j=1}^{n-1} B_j(\infty) \stackrel{p}{\to} 0 \quad \text{and} \quad \sum_{j=1}^{n-1} B_j(\infty) \stackrel{d}{\to} Po(\lambda) \quad \text{as } \theta \log n \to \lambda ,$$

respectively. Applying Slutsky's theorem to the above two convergences, we have

$$\sum_{j=1}^{n-1} B_j \stackrel{d}{\to} Po(\lambda) \quad \text{as } \theta \log n \to \lambda .$$

Thus, by Lemma 2.2, we obtain (2.10). □

2.4 Approximate Distributions for $\mathcal{L}(K_n)$

2.4.1 Poisson Approximation

2.4.1.1 Poisson Approximation (I) (Arratia et al. [5, 6])

For the logarithmic combinatorial structure, including the ESF, Poisson approximations to $\mathcal{L}(K_n)$ have been derived by Arratia et al. ([5, Theorem 5.4, Corollary 5.5 and Remark], [6, Theorem 8.15]) in detail. We quote their approximations in the case of the ESF, following their notation. Let $p(x, \eta)$ be the value of the p.f. of the Poisson distribution $Po(\eta)$ at the point x. Let $\tau_n = \theta[\psi(n + 1) - \psi(\theta + 1)]$, $a_n = -\theta^2 \psi'(\theta + 1)$, and

$$v_n\{s + 1\} = p(s, \tau_n)\left(1 + \frac{a_n}{2\tau_n^2}\{(s - \tau_n)^2 - \tau_n\}\right),$$

where $\psi(t) = \Gamma'(t)/\Gamma(t)$ and $\psi'(t)$ are the digamma and trigamma functions, respectively. Their approximations to $\mathcal{L}(K_n)$ are as follows:

$$\nu_n, \quad Po(1 + \tau_n) \text{ and } 1 + Po(\tau_n). \tag{2.21}$$

These approximations (2.21) distribute around

$$1 + \tau_n = 1 + \theta[\psi(n+1) - \psi(\theta+1)].$$

Here, we note the relations $\psi(z+1) = \psi(z) + 1/z$, $\psi'(x) = \sum_{j=0}^{\infty} 1/(z+j)^2$ and $\psi(1) = -\gamma$, where γ is Euler's constant (e.g., see [16, 8 of 8.363, 1 of 8.365 and 1 of 8.366]). Because

$$\mu_n := E(L_n) = \sum_{i=2}^{n} \frac{\theta}{\theta+i-1} = \theta[\psi(\theta+n) - \psi(\theta+1)], \tag{2.22}$$

we have

$$E(K_n) = 1 + \mu_n = 1 + \theta[\psi(n+\theta) - \psi(\theta+1)].$$

The function ψ is a monotone increasing function, because $\psi'(x) > 0$ $(x > 0)$. Therefore, μ_n becomes larger than τ_n as θ (> 1) increases. For $0 < \theta < 1$, we have $-\theta/n < \mu_n - \tau_n < 0$, because $\psi(n) = \psi(n+1) - 1/n$. We show the curves of τ_n (dashed curve) and μ_n (solid curve) as the function θ, for $n = 25, 50$, in Figs. 2.1 and 2.2. For small θ, τ_n is almost equal to μ_n. As θ increases, τ_n becomes smaller than μ_n. Because the approximations are considered to be based on the Poisson distributions, the centers and dispersions of (2.21) are smaller than those of K_n for large θ. Thus, in the next section, we consider approximations in which the centers and dispersions are equal to $E(K_n)$.

2.4.1.2   Poisson Approximation (II) (Yamato [23])

We let

$$\mu_{2,n} = \sum_{i=2}^{n} \left(\frac{\theta}{\theta+i-1} \right)^2 = \theta^2[\psi'(\theta+1) - \psi'(\theta+n)]. \tag{2.23}$$

By shifting the approximations to $\mathcal{L}(L_n)$, we obtain the approximation to $\mathcal{L}(K_n)$, as follows:

$$1 + Po(\mu_n) \text{ and } sPo : 1 + p(m, \mu_n)\left(1 - \frac{\mu_{2,n}}{2}C_2(m, \mu_n)\right), \tag{2.24}$$

where C_2 is the second Charlier polynomial,

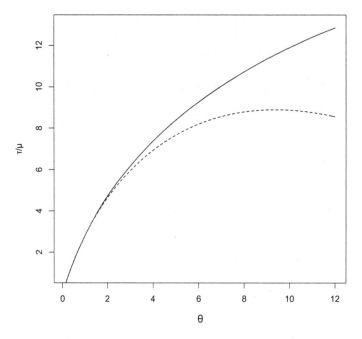

Fig. 2.1 τ_n(dashed) and μ_n(solid) for $n = 25$

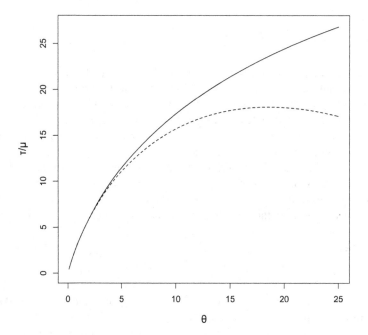

Fig. 2.2 τ_n(dashed) and μ_n(solid) for $n = 50$

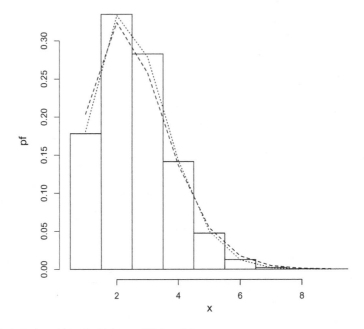

Fig. 2.3 $1+Po(\mu_n)$ (da), sPo (do); $n = 25$, $\theta = 0.5$

$$C_2(x, \lambda) = \frac{x^2 - (2\lambda + 1)x + \lambda^2}{\lambda^2}.$$

For the Charlier polynomials, see, for example, [8, p. 175], [19, 26].

The expectations and variances of (2.24) are equal to those of K_n, in contrast to (2.21). We compare the approximations $1 + Po(\mu_n)$ and sPo with the approximate p.f.s of K_n in examples based on $n = 25$ and $\theta = 0.5, 5$. In Figs. 2.3 and 2.4, the approximate p.f.s of K_n are drawn as bar graphs and are simulated using the programming language R, based on $K_n = \xi_1 + \cdots + \xi_n$. The approximations $1 + Po(\mu_n)$ and sPo are drawn as dashed and dotted lines, respectively.

The sPo is the better approximation. However, a disadvantage of the sPo is that its tails may be negative (see the right tail of sPo in Fig. 2.4). This is because sPo is a signed measure based on the first two terms of an orthogonal expansion by Charlier polynomials.

Note that

$$E(L_n) = \mu_n \text{ and } Var(L_n) = \mu_n\left(1 - \frac{\mu_{2,n}}{\mu_n}\right).$$

If $\mu_{2,n}/\mu_n$ is small, then $Var(L_n)$ is close to $E(L_n)$ and, therefore, the Poisson distribution is appropriate for the approximation to $\mathcal{L}(L_n)$. In general, because $E(L_n) > Var(L_n)$, the binomial distribution may be more appropriate for the approximation to $\mathcal{L}(L_n)$. Therefore, we consider this distribution for the approximation, obtaining the shifted binomial approximation to $\mathcal{L}(K_n)$.

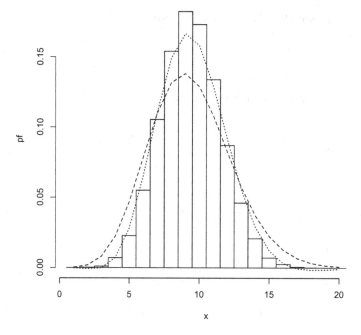

Fig. 2.4 $1+\text{Po}(\mu_n)$ (da), sPo (do); $n = 25, \theta = 5$

2.4.2 Binomial Approximation

2.4.2.1 Approximation I (Yamato [24, 25])

Let $g(x; n, p)$ be the p.f. of the binomial distribution $B_N(n, p)$. For the sum S_n of independent Bernoulli variables, [8, p. 190] propose a binomial approximation with mean equal to $E(S_n)$ and variance approximately equal to $V(S_n)$.

Using this method to obtain L_n, we have

$$(n - 1)' = \lfloor \mu_n^2 / \mu_{2,n} \rfloor \text{ and } p' = \mu_n / (n - 1)' ,$$

where $\lfloor x \rfloor$ is an integer close to x. Then, we have the following shifted binomial approximation to $\mathcal{L}(K_n)$:

$$\text{I} : 1 + B_N((n - 1)', p') . \tag{2.25}$$

The approximate p.f. of K_n is given by

$$g(x - 1; (n - 1)', p') \quad (x = 1, 2, \ldots, (n - 1)' + 1) . \tag{2.26}$$

2.4.2.2 Approximation II (Yamato [24, 25])

We put

$$\bar{p}_{n-1} = \mu_n/(n-1), \quad \gamma_2(\bar{p}_{n-1}) = \mu_{2,n} - (n-1)\bar{p}_{n-1}^2 .$$

Here, we use a Krawtchouk polynomial. Put

$$\mathcal{B}_2(n-1, \bar{p}_{n-1})(\{x\}) = g(x; n-1, \bar{p}_{n-1}) - \frac{\gamma_2(\bar{p}_{n-1})}{2} \Delta^2 g(x; n-3, \bar{p}_{n-1}) ,$$

$$\Delta^2 g(x; n-3, p) = \frac{g(x; n-1, p)}{(n-1)(n-2)\,p^2\,(1-p)^2} \left\{ x^2 - [1 + 2(n-2)\,p]x + (n-1)(n-2)\,p^2 \right\} ,$$

where $n^{(2)} \Delta^2 g(x, n-2, p)/g(x, n, p)$ is the second Krawtchouk polynomial; for further information, see, for example, [18, 20]. As an approximation to $\mathcal{L}(K_n)$, we obtain the shifted finite-signed measure, as follows:

$$\text{II} : \ 1 + \mathcal{B}_2(n-1, \bar{p}_{n-1})(\{x\}) , \tag{2.27}$$

which has the same first two moments as K_n. Both (2.26), ((2.25)) and (2.27) perform well, as shown by the dashed and dotted lines in Figs. 2.5 and 2.6, respectively. However, the tails of (2.27) may be negative, albeit small. This occurs because (2.27) is a signed measure based on the first two terms of an orthogonal expansion, similarly to sPo in (2.24). For example, for $n = 25$ and $\theta = 0.5$, the right tail of (2.27) has small negative values -3.6×10^{-6} and -9.0×10^{-7}, for $x = 9$ and 10, respectively. For $n = 25$ and $\theta = 5$, the right tail of (2.27) has small negative values -6.6×10^{-6} and -3.9×10^{-6}, for $x = 18$ and 19, respectively. Therefore, the corresponding distribution function is not monotone increasing. Thus, we recommend using the shifted binomial approximation I (2.26), ((2.25)) as the approximation to $\mathcal{L}(K_n)$. We discuss its two applications of this approximation [25] in the next section.

2.5 Applications

2.5.1 An Approximation to the p.f. of the MLE $\hat{\theta}$ of θ

Given observation $K_n = k$, the MLE $\hat{\theta}$ of the parameter θ is the solution of the following equation [12]:

$$k = \sum_{j=1}^{n} \frac{\theta}{\theta + j - 1} . \tag{2.28}$$

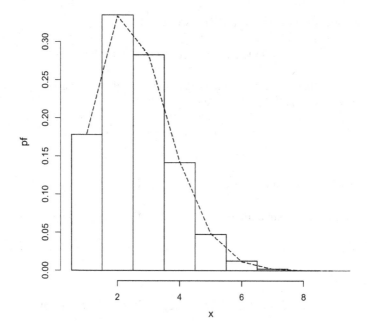

Fig. 2.5 I (dash) and II (dot); $n = 25$, $\theta = 0.5$

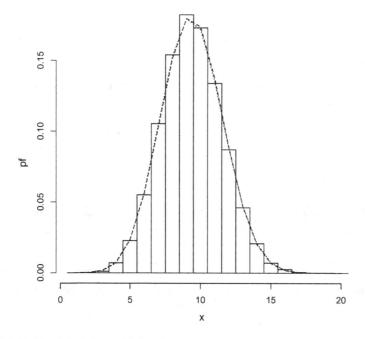

Fig. 2.6 I (dash) and II (dot); $n = 25$, $\theta = 5$

Using the digamma function ψ, (2.28) is written as

$$k = \mu_n(\theta), \quad \mu_n(\theta) = \theta[\psi(\theta + n) - \psi(\theta)] .$$

Because $\mu_n(\theta)$ is a strictly increasing function of θ, for each $k = 1, 2, \ldots, n$, there exists a unique $\mu_n^{-1}(k)$. Thus, we have

$$P(\hat{\theta} = \mu_n^{-1}(k)) = P(K_n = k) \quad (k = 1, 2, \ldots, n) ,$$

or

$$P(\hat{\theta} = x) = P(K_n = \mu_n(x)) \quad (x = \mu_n^{-1}(k), \ k = 1, 2, \ldots, n) .$$

Using (2.26), the approximation to the p.f. of the MLE $\hat{\theta}$ is given by

$$P(\hat{\theta} = x) \doteqdot g(k - 1; (n - 1)', p') \quad (\mu_n(x) = k, \ k = 1, 2, \ldots, n) .$$

2.5.2 An Estimation of the p.f. of K_n When θ Is Unknown

The necessary values for the approximation (2.25), (2.26) are

$$\lambda_{n-1} = \sum_{i=2}^{n} \frac{\theta}{\theta + i - 1}, \quad \lambda_{2,n-1} = \sum_{i=2}^{n} \left(\frac{\theta}{\theta + i - 1}\right)^2 = \theta^2[\psi'(\theta + 1) - \psi'(\theta + n)] .$$

As an estimator of θ, we use the MLE $\hat{\theta}$, which satisfies $k = \sum_{j=1}^{n} \theta/(\theta + j - 1)$. Then, from the above relations, we have

$$\lambda_{n-1} = k - 1, \quad \lambda_{2,n-1}^{**} := \lambda_{2,n-1} = \hat{\theta}^2[\psi'(\hat{\theta} + 1) - \psi'(\hat{\theta} + n)] .$$

We put

$$(n - 1)^{**} = \lfloor (k - 1)^2/\lambda_{2,n-1}^{**} \rfloor, \quad p^{**} = (k - 1)/(n - 1)^{**} .$$

Using the approximation given in (2.26), we obtain the estimator of the p.f. of K_n, as follows:

$$g(x - 1; (n - 1)^{**}, p^{**}) \quad (x = 1, 2, \ldots, (n - 1)^{**} + 1) .$$

Acknowledgements The author thanks the anonymous referee and the editor for their careful reading of the manuscript and their valuable feedback.

Appendix

We explain the three relations used in the introduction.

We first show (2.4) for **Chinese restaurant process (I)**. For each $j = 1, 2, \ldots, k$, only the first customer of the jth table associated with A_j is specified. Therefore, for positive integers $a_1, a_2, \ldots, a_k$ satisfying $a_1 + a_2 + \cdots + a_k = n$, we have

$$
P(A_1 = a_1, A_2 = a_2, \ldots, A_k = a_k)
$$
$$
= \frac{\theta^k}{\theta^{[n]}} \times P(n-1, a_1-1) P(n-a_1-1, a_2-1) \cdots P(n-a_1 - \cdots - a_{k-1}-1, a_k-1) \cdot (a_k-1)!
$$
$$
= \frac{\theta^k}{\theta^{[n]}} \times \frac{(n-1)!}{(n-a_1)(n-a_1-a_2)\cdots(n-a_1-a_2-\cdots-a_{k-1})} , \tag{2.29}
$$

where $P(m, r)$ is an r-permutation of m given by $P(m, r) = m!/(m-r)!$. The last term of (2.29) is equal to the right-hand side of (2.4).

We next show (2.8) for **Chinese restaurant process (II)**. The ordered spacing $(D_1, D_2, \ldots, D_k)$ is based on the $0-1$ sequence $\xi_1 \xi_2 \cdots \xi_n 1$ ($\xi_1 = 1$). Its reversed sequence $1 \xi_n \xi_{n-1} \cdots \xi_1$ gives the reversed spacing $(D_k, D_{k-1}, \ldots, D_1)$, and is associated with the Feller coupling (e.g., see [4, Sect. 3]). Therefore, we note the following relation:

$$
P(D_1 = d_1, D_2 = d_2, \ldots, D_k = d_k) = P(D_k = d_k, D_{k-1} = d_{k-1}, \ldots, D_1 = d_1). \tag{2.30}
$$

Let $H(\theta, m, 1) := \theta/(\theta + m - 1)$ and

$$
H(\theta, m, d) := \frac{m-1}{\theta+m-1} \cdots \frac{m-(d-1)}{\theta+m-(d-1)} \cdot \frac{\theta}{\theta+m-d} \quad (d > 1) .
$$

We apply the property of the Feller coupling to the right-hand side of (2.30) using the above H. Then, for positive integers $d_1, d_2, \ldots, d_k$ satisfying $d_1 + d_2 + \cdots + d_k = n$, we have

$$
P(D_k = d_k, D_{k-1} = d_{k-1}, \ldots, D_1 = d_1)
$$
$$
= H(\theta, n, d_k) \cdot H(\theta, n-d_k, d_{k-1}) \cdots H(\theta, n-(d_k + \cdots + d_2), d_1) , \tag{2.31}
$$

By (2.30) and (2.31), we have

$$
P(D_1 = d_1, D_2 = d_2, \ldots, D_k = d_k)
$$
$$
= \frac{\theta^k}{\theta^{[n]}} (n-1) \cdots (n-(d_k-1)) \times (n-(d_k+1)) \cdots (n-(d_k+d_{k-1}-1)) \times (n-(d_k+d_{k-1}+1))
$$
$$
\cdots (n-(d_k+d_{k-1}+\cdots+d_2-1)) \times (n-(d_k+d_{k-1}+\cdots+d_2+1)) \cdots (n-(d_k+\cdots+d_2+d_1-1))
$$

$$= \frac{\theta^k}{\theta^{[n]}} \frac{(n-1)!}{(n-d_k)(n-d_k-d_{k-1})\cdots(n-d_k-d_{k-1}-\cdots-d_2)},\qquad(2.32)$$

which is equal to the right-hand side of (2.8).

Finally, we consider the relation connecting (2.4) and (2.8) to the ESF.

Proposition 2.2 (Donnelly and Tavaraé [10, Proposition 2.1]) *Let* $\mu_1, \mu_2, \ldots, \mu_l$ *be any l positive integers, and let Π be the collection of permutations $\pi = \big(\pi(1), \ldots, \pi(l)\big)$ of $(1, 2, \ldots, l)$. Then,*

$$\sum_{\pi \in \Pi} \frac{1}{\mu_{\pi(1)}(\mu_{\pi(1)} + \mu_{\pi(2)}) \cdots (\mu_{\pi(1)} + \mu_{\pi(2)} + \cdots + \mu_{\pi(l)})} = \frac{1}{\mu_1 \mu_2 \cdots \mu_l}.$$

For this relation, we consider the case in which duplications exist among $\mu_1, \mu_2, \ldots, \mu_l$, and let $b_1, \ldots, b_k$ be their distinct values. Let c_j be the number of μ_i equal to b_j, for $j = 1, \ldots, k$ and $i = 1, \ldots, l$.

Let Π^* be the collection of permutations $\pi = \big(\pi(1), \ldots, \pi(l)\big)$, excluding the duplications of $\mu_1, \mu_2, \ldots, \mu_l$. Then, we have the following corollary:

Corollary 2.1

$$\sum_{\pi \in \Pi^*} \frac{1}{\mu_{\pi(1)}(\mu_{\pi(1)} + \mu_{\pi(2)}) \cdots (\mu_{\pi(1)} + \mu_{\pi(2)} + \cdots + \mu_{\pi(l)})} = \frac{1}{\mu_1 \cdots \mu_l c_1! \cdots c_k!}$$

$$= \prod_{j=1}^{k} \frac{1}{b_j{}^{c_j} c_j!}. \qquad (2.33)$$

References

1. Aldous DJ (1985) Exchangeability and related topics, Ecole d'Etéde Probabilités de Saint-Flour XIII 1983, vol 1117. Lecture Notes in Mathematics. Springer, Berlin, pp 1–198
2. Antoniak CE (1974) Mixtures of Dirichlet processes with applications to Bayesian nonparametric problems. Ann Stat 2:1142–1174
3. Arratia R, Tavaré S (1992) Limit theorems for combinatorial structures via discrete process approximations. Random Struct Algorithms 3(3):321–345
4. Arratia R, Barbour AD, Tavaré S (1992) Poisson processes approximations for the Ewens sampling formula. Ann Appl Probab 2:519–535
5. Arratia R, Barbour AD, Tavaré S (2000) The number of components in logarithmic combinatorial structure. Ann Appl Probab 10:331–361
6. Arratia R, Barbour AD, Tavaré S (2003) Logarithmic combinatorial structures: a probabilistic approach. EMS monographs in mathematics. EMS Publishing House, Zürich
7. Arratia R, Barbour AD, Tavaré S (2016) Exploiting the Feller coupling for the Ewens sampling formula [comment on Crane (2016)]. Stat Sci 31(1):27–29
8. Barbour AD, Holst L, Janson S (1992) Poisson approximation. Clarendon Press, Oxford
9. Crane H (2016) The ubiquitous Ewens sampling formula. Stat Sci 31(1):1–19
10. Donnelly P, Tavaraé S (1986) The ages of alleles and coalescent. Adv Appl Probab 18:1–19

11. Ethier SN (1990) The infinitely-many-neutral-alleles diffusion model with ages. Adv Appl Probab 22:1–24
12. Ewens WJ (1972) The sampling theory of selectively neutral alleles. Theor Popul Biol 3:87–112
13. Ewens WJ (1990) Population genetics theory - the past and the future. In: Lessard S (ed) Mathematical and statistical developments of evolutionary theory. Kluwer, Amsterdam, pp 177–227
14. Ferguson TS (1973) A Bayesian analysis of some nonparametric problems. Ann Stat 1:209–230
15. Freedman D (1974) The Poisson approximation for dependent eventst. Ann Probab 2:256–269
16. Gradshteyn IS, Ryzhik IM (2007) In: Jeffrey A, Zwillinger D (eds) Table of integrals, series, and products. Academic Press
17. Johnson NL, Kotz S, Balakrishnan N (1997) Discrete multivariate distributions. Wiley, New York
18. Roos B (2001) Binomial approximation to the Poisson binomial distribution: the Krawtchouk expansion. Theory Probab Appl 45:258–272
19. Takeuchi K (1975) Approximation of probability distributions (in Japanese). Kyouiku Shuppan, Tokyo
20. Takeuchi K, Takemura A (1987) On sum of 0–1 random variables I. univariate case. Ann Inst Stat Math 39:85–102
21. Wang YH (1993) On the number of successes in independent trials. Statistica Sinica 3:295–312
22. Yamato H (1997) On the Donnelly-Tavaré-Griffiths formula associated with the coalescent. Commun Stat-Theory Methods 26:589–599
23. Yamato H (2017a) Poisson approximations for sum of Bernoulli random variables and its application to Ewens sampling formula. J Jpn Stat Soc 47(2):187–195
24. Yamato H (2017b) Shifted binomial approximation for the Ewens sampling formula. Bull Inform Cybern 49:81–88
25. Yamato H (2018) Shifted binomial approximation for the Ewens sampling formula (II). Bull Inform Cybern 50:43–50
26. Zacharovas V, Hwang H-K (2010) A Charlier-Parseval approach to Poisson approximation and its applications. Lith Math J 50:88–119

Chapter 3
Error Bounds for the Normal Approximation to the Length of a Ewens Partition

Koji Tsukuda

Abstract Let $K(= K_{n,\theta})$ be a positive integer-valued random variable with a distribution given by $P(K = x) = \bar{s}(n, x)\theta^x/(\theta)_n$ $(x = 1, \ldots, n)$, where θ is a positive value, n is a positive integer, $(\theta)_n = \theta(\theta + 1) \cdots (\theta + n - 1)$, and $\bar{s}(n, x)$ is the coefficient of θ^x in $(\theta)_n$ for $x = 1, \ldots, n$. This formula describes the distribution of the length of a Ewens partition, which is a standard model of random partitions. As n tends to infinity, K asymptotically follows a normal distribution. Moreover, as n and θ simultaneously tend to infinity, if $n^2/\theta \to \infty$, K also asymptotically follows a normal distribution. This study provides error bounds for the normal approximation. The results show that the decay rate of the error varies with asymptotic regimes.

Keywords Berry–Esseen-type theorem · Error bound · Ewens sampling formula · Infinitely many neutral allele model · Normal approximation · Random partition

3.1 Introduction

Consider a nonnegative integer-valued random variable $K(= K_{n,\theta})$ that follows

$$P(K = x) = \frac{\bar{s}(n, x)\theta^x}{(\theta)_n} \quad (x = 1, \ldots, n), \tag{3.1}$$

where θ is a positive value, n is a positive integer, $(\theta)_n = \theta(\theta + 1) \cdots (\theta + n - 1)$, and $\bar{s}(n, x)$ is the coefficient of θ^x in $(\theta)_n$. This distribution is known as the falling factorial distribution [18, Eq. (2.22)], STR1F (i.e., the Stirling family of distributions with finite support related to the Stirling number of the first kind) [13, 14], and the Ewens distribution [12]. The formula (3.1) describes the distribution of the length of a Ewens partition, which is a standard model of random partitions. A

K. Tsukuda (✉)
Faculty of Mathematics, Kyushu University, 744 Motooka, Nishi-ku, Fukuoka-shi, Fukuoka 819-0395, Japan
e-mail: tsukuda@math.kyushu-u.ac.jp

© The Author(s), under exclusive license to Springer Nature Singapore Pte Ltd. 2020   55
N. Hoshino et al. (eds.), *Pioneering Works on Distribution Theory*,
JSS Research Series in Statistics,
https://doi.org/10.1007/978-981-15-9663-6_3

random partition is called a Ewens partition if its distribution is given by the Ewens sampling formula. This formula and (3.1) appear in many scientific fields, and thus have been studied extensively; see, for example, Johnson et al. [11, Chap. 41] or Crane [3]. In the context of population genetics, (3.1) was discussed in Ewens [5] as the distribution of the number of allelic types in a sample of size n generated from the infinitely many neutral allele model with scaled mutation rate θ; see also Durrett [4, Sect. 1.3]. Moreover, in the context of nonparametric Bayesian inference, (3.1) describes the law of the number of distinct values in a sample from the Dirichlet process; see, for example, Ghosal and van der Vaart [7, Sect. 4.1]. Furthermore, as noted in Sibuya [13], (3.1) relates to several statistical and combinatorial topics, including permutations, sequential rank-order statistics, and binary search trees.

Simple calculations imply that

$$E[K] = \theta \sum_{i=1}^{n} \frac{1}{\theta + i - 1}, \quad \text{var}(K) = \theta \sum_{i=1}^{n} \frac{i-1}{(\theta + i - 1)^2},$$

and

$$E[K] \sim \text{var}(K) \sim \theta \log n \tag{3.2}$$

as $n \to \infty$. Let $\tilde{F}_{n,\theta}(\cdot)$ be the distribution function of the random variable

$$Z_{n,\theta} = \frac{K - \theta \log n}{\sqrt{\theta \log n}}$$

standardized by the leading terms of the mean and variance, and let $\Phi(\cdot)$ be the distribution function of the standard normal distribution. By calculating the moment generating function of $Z_{n,\theta}$, Watterson [19] proved that $Z_{n,\theta}$ converges in distribution to the standard normal distribution; that is, $\tilde{F}_{n,\theta}(x) \to \Phi(x)$ as $n \to \infty$ for any $x \in \mathbb{R}$; for the history concerning this result, refer to Arratia and Tavaré [1, Remark after Theorem 3]. In particular, when $\theta = 1$, Goncharov [8] proved that $\tilde{F}_{n,1}(x) \to \Phi(x)$ for any $x \in \mathbb{R}$. From a theoretical perspective, it is important to derive error bounds for the approximation. Yamato [20] investigated the first-order Edgeworth expansion of $\tilde{F}_{n,\theta}(\cdot)$ using a Poisson approximation [1, Remark after Theorem 3] and proved that $\|\tilde{F}_{n,\theta} - \Phi\|_\infty = O\left(1/\sqrt{\log n}\right)$, where $\|\cdot\|_\infty$ is the ℓ^∞-norm defined by

$$\|f\|_\infty = \sup_{x \in \mathbb{R}} |f(x)|$$

for a bounded function $f : \mathbb{R} \ni x \mapsto f(x)$. Note that when $\theta = 1$, Hwang [10, Example 1] showed that $\|\tilde{F}_{n,1} - \Phi\|_\infty = O\left(1/\sqrt{\log n}\right)$. Kabluchko et al. [12] derived the Edgeworth expansion of the probability function of K and the first-order Edgeworth expansion of $\tilde{F}_{n,\theta}(\cdot)$.

Because the standardization of $Z_{n,\theta}$ comes from (3.2), the normal approximation only works well when n is sufficiently large compared with θ. However, this assumption has limited validity in practical cases. Therefore, it is important to consider alternative standardized variables; see, for example, Yamato [20] and Yamato et al. [21]. In particular, we consider the random variables $X_{n,\theta}$ and $Y_{n,\theta}$ defined by

$$X_{n,\theta} = \frac{K - \mu_0}{\sigma_0} \quad \text{and} \quad Y_{n,\theta} = \frac{K - \mu_T}{\sigma_T},$$

respectively, where

$$\mu_0 = \mathrm{E}[K], \quad \sigma_0^2 = \mathrm{var}(K),$$

$$\mu_T = \theta \log\left(1 + \frac{n}{\theta}\right), \quad \text{and} \quad \sigma_T^2 = \theta\left(\log\left(1 + \frac{n}{\theta}\right) + \frac{\theta}{n+\theta} - 1\right).$$

These are standardized random variables that use the exact moments and approximate moments, respectively. Denote the distribution functions of $X_{n,\theta}$ and $Y_{n,\theta}$ by $F_{n,\theta}(\cdot)$ and $G_{n,\theta}(\cdot)$, respectively. Then, Tsukuda [15, Theorem 2 and Remark 6] proved that, under the asymptotic regime $n^2/\theta \to \infty$ and $\theta \not\to 0$ as $n \to \infty$ (see Sect. 3.1.1 for the explicit assumptions), both $F_{n,\theta}(x)$ and $G_{n,\theta}(x)$ converge to $\Phi(x)$ as $n \to \infty$ for any $x \in \mathbb{R}$. This study provides upper and lower bounds for the approximation errors $\|F_{n,\theta} - \Phi\|_\infty$ and $\|G_{n,\theta} - \Phi\|_\infty$.

Remark 3.1 It holds that $\mu_0 \sim \mu_T$ and $\sigma_0 \sim \sigma_T$ as $n \to \infty$ with $n^2/\theta \to \infty$.

3.1.1 Assumptions and Asymptotic Regimes

As explained in the introduction, the regime $n \to \infty$ with fixed θ is sometimes unrealistic. Hence, we consider asymptotic regimes in which θ increases as n increases. Such regimes have been discussed in Feng [6, Sect. 4] and Tsukuda [15, 16]. We follow these studies. In this subsection, we summarize the assumptions on n and θ.

First, θ is assumed to be nondecreasing with respect to n. Moreover, when we take the limit operation, $n^2/\theta \to \infty$ is assumed.

The following asymptotic regimes are discussed in this paper:

- Case A: $n/\theta \to \infty$
- Case B: $n/\theta \to c$, where $0 < c < \infty$
- Case C: $n/\theta \to 0$
- Case C1: $n/\theta \to 0$ and $n^2/\theta \to \infty$

Remark 3.2 Feng [6] was apparently the first to consider asymptotic regimes in which n and θ simultaneously tend to infinity. Specifically, Feng [6, Sect. 4] considered Cases A, B, and C. Case C1 was introduced by Tsukuda [15].

Furthermore, let $c^\star$ be the unique positive root of the equation

$$\log(1+x) - 2 + \frac{3}{x+1} - \frac{1}{(x+1)^2} = 0. \tag{3.3}$$

Then, we introduce a new regime, Case B*, as follows.

- Case B*: $n/\theta \to c$, where $0 < c < \infty$ and $c \neq c^*$.

Remark 3.3 Solving (3.3) numerically gives $c^* = 2.16258\cdots$.

3.2 Main Results

This section presents Theorems 3.1 and 3.2, which are the main results of this study, as well as their corollaries. Proofs of the results in this section are provided in Sect. 3.4.

3.2.1 An Upper Error Bound

In this subsection, an upper bound for the error $\|F_{n,\theta} - \Phi\|_\infty$ is given in Theorem 3.1, and its convergence rate is given in Corollary 3.1. Moreover, the convergence rate of the upper bound for the error $\|G_{n,\theta} - \Phi\|_\infty$ is given in Corollary 3.2.

We now present the first main theorem of this paper.

Theorem 3.1 *Assume that there exists $n_0(= n_0(\theta))$ such that*

$$\theta \left(\log \left(1 + \frac{n}{\theta} \right) - 1 + \frac{\theta}{n+\theta} \right) + \frac{n}{2(\theta + n)} - 1 > 0 \tag{3.4}$$

for all $n \geq n_0$. Then, it holds that

$$\|F_{n,\theta} - \Phi\|_\infty \leq C\gamma_1$$

for all $n \geq n_0$, where C is a constant not larger than 0.5591, and

$$\gamma_1 = \frac{\theta \left\{ \log \left(1 + \frac{n}{\theta} \right) - \frac{5}{3} + \frac{3\theta}{n+\theta} - \frac{2\theta^2}{(n+\theta)^2} + \frac{2\theta^3}{3(n+\theta)^3} \right\} + 4 + \frac{n}{n+\theta}}{\left\{ \theta \left(\log \left(1 + \frac{n}{\theta} \right) - 1 + \frac{\theta}{n+\theta} \right) + \frac{n}{2(\theta+n)} - 1 \right\}^{3/2}}. \tag{3.5}$$

Remark 3.4 Under our asymptotic regime $(n^2/\theta \to \infty)$, (3.4) is valid for sufficiently large n.

Remark 3.5 The constant C in Theorem 3.1 is the universal constant in the Berry–Esseen theorem.

Theorem 3.1 and the asymptotic evaluations of the numerator and denominator of γ_1 yield the following corollary.

Corollary 3.1 *In Cases A, B, and C1, it holds that*

$$\|F_{n,\theta} - \Phi\|_\infty = O\left(\frac{1}{\sqrt{\theta\{\log(1 + n/\theta) - 1 + \theta/(n + \theta)\}}}\right)$$

$$= \begin{cases} O\left(1/\sqrt{\theta \log(n/\theta)}\right) & \text{(Case A)}, \\ O\left(1/\sqrt{\theta}\right) & \text{(Case B)}, \\ O\left(1/\sqrt{n^2/\theta}\right) & \text{(Case C1)}. \end{cases}$$

Using Corollary 3.1, we can obtain the following convergence rate of the error bound for the normal approximation to $Y_{n,\theta}$.

Corollary 3.2 *It holds that*

$$\|G_{n,\theta} - \Phi\|_\infty = \begin{cases} O\left(1/\sqrt{\theta \log(n/\theta)}\right) & \text{(Case A)}, \\ O\left(1/\sqrt{\theta}\right) & \text{(Case B)}, \\ O\left(1/\sqrt{n^2/\theta}\right) & \text{(Case C1)}. \end{cases}$$

3.2.2 Evaluation of the Decay Rate

In this subsection, a lower bound for the error $\|F_{n,\theta} - \Phi\|_\infty$ is given in Theorem 3.2. Together with Theorem 3.1, this theorem yields the decay rate of $\|F_{n,\theta} - \Phi\|_\infty$, as stated in Corollary 3.3.

We now present the second main theorem of this paper.

Theorem 3.2 *(i) Assume that there exists $n_1(= n_1(\theta))$ such that, for all $n \geq n_1$, (3.4) holds, $\mathrm{var}(K) \geq 1$, and*

$$\theta\left\{\log\left(1 + \frac{n}{\theta}\right) - 2 + \frac{3\theta}{n + \theta} - \frac{\theta^2}{(n + \theta)^2}\right\} - 3 + \frac{n}{2(n + \theta)} > 0. \tag{3.6}$$

Then, it holds that

$$\|F_{n,\theta} - \Phi\|_\infty \geq \frac{\gamma_2}{D} - \gamma_3$$

for all $n \geq n_1$, where D is some constant,

$$\gamma_2 = \frac{\theta\left\{\log\left(1 + \frac{n}{\theta}\right) - 2 + \frac{3\theta}{n+\theta} - \frac{\theta^2}{(n+\theta)^2}\right\} - 3 + \frac{n}{2(n+\theta)}}{\{\theta\left(\log\left(1 + \frac{n}{\theta}\right) - 1 + \frac{\theta}{n+\theta}\right) + \frac{n}{\theta+n}\}^{3/2}}, \tag{3.7}$$

and

$$\gamma_3 = \frac{\theta \left\{ \frac{1}{3} - \frac{\theta}{n+\theta} + \frac{\theta^2}{(n+\theta)^2} - \frac{\theta^3}{3(n+\theta)^3} \right\} + 2}{\left\{ \theta \left(\log \left(1 + \frac{n}{\theta} \right) - 1 + \frac{\theta}{n+\theta} \right) + \frac{n}{2(\theta+n)} - 1 \right\}^2}. \tag{3.8}$$

(ii) *Assume that there exists $n_2(= n_2(\theta))$ such that, for all $n \geq n_2$, (3.4) holds,* var$(K) \geq 1$, *and*

$$\theta \left\{ \log \left(1 + \frac{n}{\theta} \right) - 2 + \frac{3\theta}{n+\theta} - \frac{\theta^2}{(n+\theta)^2} \right\} + 2 + \frac{n}{n+\theta} < 0. \tag{3.9}$$

Then, it holds that

$$\| F_{n,\theta} - \Phi \|_\infty \geq \frac{\gamma_4}{D} - \gamma_3$$

for all $n \geq n_2$, where D is some constant, γ_3 is as defined in (3.8), and

$$\gamma_4 = \frac{-\left[\theta \left\{ \log \left(1 + \frac{n}{\theta} \right) - 2 + \frac{3\theta}{n+\theta} - \frac{\theta^2}{(n+\theta)^2} \right\} + 2 + \frac{n}{n+\theta} \right]}{\left\{ \theta \left(\log \left(1 + \frac{n}{\theta} \right) - 1 + \frac{\theta}{n+\theta} \right) + \frac{n}{\theta+n} \right\}^{3/2}}. \tag{3.10}$$

Remark 3.6 Under our asymptotic regime ($n^2/\theta \to \infty$), var$(K) \geq 1$ is valid for sufficiently large n. In Case A, (3.6) is valid for sufficiently large n. In Case B*, if $c > c^\star$ then (3.6) is valid for sufficiently large n, and if $c < c^\star$ then (3.9) is valid for sufficiently large n. In Case C1, (3.9) is valid for sufficiently large n.

Remark 3.7 The constant D in Theorem 3.2 is the universal constant introduced by Hall and Barbour [9]. Note that they denote this constant as C in their theorem.

As a corollary to Theorems 3.1 and 3.2, we can make the following statement regarding the decay rate of $\| F_{n,\theta} - \Phi \|_\infty$.

Corollary 3.3 *It holds that*

$$\| F_{n,\theta} - \Phi \|_\infty \asymp \begin{cases} 1/\sqrt{\theta} \log (n/\theta) & \text{(Case A)}, \\ 1/\sqrt{\theta} & \text{(Case B}^\star), \\ 1/\sqrt{n^2/\theta} & \text{(Case C1)}. \end{cases}$$

3.3 Preliminary Results

3.3.1 A Representation of K Using a Bernoulli Sequence

Consider an independent Bernoulli random sequence $\{\xi_i\}_{i \geq 1} (= \{\xi_{i,\theta}\}_{i \geq 1})$ defined by

$$P(\xi_i = 1) = p_i = \frac{\theta}{\theta + i - 1}, \quad P(\xi_i = 0) = 1 - p_i \quad (i = 1, 2, \ldots).$$

Then,

$$P(K = x) = P\left(\sum_{i=1}^{n} \xi_i = x\right) \quad (x = 1, \ldots, n); \tag{3.11}$$

that is, $\mathcal{L}(K)$ is equal to $\mathcal{L}(\sum_{i=1}^{n} \xi_i)$; see, for example, Johnson et al. [11, Eq. (41.12)] or Sibuya [13, Proposition 2.1]. By virtue of this relation, and after some preparation, we prove the results presented in Sect. 3.2. To use the Berry–Esseen-type theorem for independent random sequences (see Lemma 3.7), we evaluate the sum of the second- and third-order absolute central moments of $\{\xi_i\}_{i=1}^{n}$. That is, we evaluate

$$\sum_{i=1}^{n} E[|\xi_i - p_i|^2] = \theta \sum_{i=1}^{n} \frac{1}{\theta + i - 1} - \theta^2 \sum_{i=1}^{n} \frac{1}{(\theta + i - 1)^2} \tag{3.12}$$

and

$$\sum_{i=1}^{n} E[|\xi_i - p_i|^3]$$

$$= \theta \sum_{i=1}^{n} \frac{(i - 1)\{\theta^2 + (i - 1)^2\}}{(\theta + i - 1)^4}$$

$$= \theta^3 \sum_{i=1}^{n} \frac{i - 1}{(\theta + i - 1)^4} + \theta \sum_{i=1}^{n} \frac{(i - 1)^3}{(\theta + i - 1)^4}$$

$$= \theta \sum_{i=1}^{n} \frac{1}{\theta + i - 1} - 3\theta^2 \sum_{i=1}^{n} \frac{1}{(\theta + i - 1)^2} + 4\theta^3 \sum_{i=1}^{n} \frac{1}{(\theta + i - 1)^3}$$

$$- 2\theta^4 \sum_{i=1}^{n} \frac{1}{(\theta + i - 1)^4}. \tag{3.13}$$

To derive a lower bound result, we evaluate

$$\sum_{i=1}^{n} E[(\xi_i - p_i)^3] = \theta \sum_{i=1}^{n} \frac{1}{\theta + i - 1} - 3\theta^2 \sum_{i=1}^{n} \frac{1}{(\theta + i - 1)^2} + 2\theta^3 \sum_{i=1}^{n} \frac{1}{(\theta + i - 1)^3} \tag{3.14}$$

and

$$\sum_{i=1}^{n} \left(E[|\xi_i - p_i|^2]\right)^2$$

$$= \theta^2 \sum_{i=1}^{n} \frac{(i-1)^2}{(\theta+i-1)^4}$$

$$= \theta^2 \sum_{i=1}^{n} \frac{1}{(\theta+i-1)^2} - 2\theta^3 \sum_{i=1}^{n} \frac{1}{(\theta+i-1)^3} + \theta^4 \sum_{i=1}^{n} \frac{1}{(\theta+i-1)^4}. \quad (3.15)$$

Remark 3.8 It follows from the binomial theorem that

$$\sum_{i=1}^{n} E\left[(\xi_i - p_i)^m\right]$$

$$= \sum_{j=1}^{m-1} (-1)^{j-1} \binom{m}{j-1} \sum_{i=1}^{n} \left(\frac{\theta}{\theta+i-1}\right)^j + (-1)^{m-1}(m-1) \sum_{i=1}^{n} \left(\frac{\theta}{\theta+i-1}\right)^m$$

for any $m = 2, 3, \ldots$.

3.3.2   Evaluations of Moments

In this subsection, we evaluate several sums of moments of $\{\xi_i\}_{i=1}^{n}$.

Lemma 3.1 *(i) It holds that*

$$\theta \left(\log \left(1 + \frac{n}{\theta} \right) - 1 + \frac{\theta}{n+\theta} \right) + \frac{n}{2(\theta+n)} - 1$$

$$\leq \sum_{i=1}^{n} E[|\xi_i - p_i|^2]$$

$$\leq \theta \left(\log \left(1 + \frac{n}{\theta} \right) - 1 + \frac{\theta}{n+\theta} \right) + \frac{n}{\theta+n}.$$

(ii) If $n^2/\theta \to \infty$ then it holds that

$$\sum_{i=1}^{n} E[|\xi_i - p_i|^2] \sim \theta \left(\log \left(1 + \frac{n}{\theta} \right) - 1 + \frac{\theta}{n+\theta} \right).$$

(iii) In particular, it holds that

$$\sum_{i=1}^{n} E[|\xi_i - p_i|^2] \sim \begin{cases} \theta \log \left(\frac{n}{\theta} \right) & \text{(Case A)}, \\ \theta \left(\log (1+c) - 1 + \frac{1}{c+1} \right) & \text{(Case B)}, \\ \frac{n^2}{2\theta} & \text{(Case C)}. \end{cases}$$

Proof (i) The desired inequality is an immediate consequence of (3.12) and Lemma 3.5.

(ii) Because

$$\log(1+x) - 1 + \frac{1}{x+1} > 0$$

for any $x > 0$, it holds that

$$\theta\left\{\log\left(1+\frac{n}{\theta}\right) - 1 + \frac{\theta}{n+\theta}\right\} \to \infty,$$

whereas the remainder terms in the inequality of (i) does not diverge to $\pm\infty$. This implies assertion (ii).

(iii) Assertion (iii) is a direct consequence of (ii) (for Case C, the result follows from the Taylor expansion of $\log(1+x) - 1 + 1/(x+1)$ as $x \to 0$).

Lemma 3.2 *(i) It holds that*

$$\theta\left\{\log\left(1+\frac{n}{\theta}\right) - \frac{5}{3} + \frac{3\theta}{n+\theta} - \frac{2\theta^2}{(n+\theta)^2} + \frac{2\theta^3}{3(n+\theta)^3}\right\} + \frac{n}{2(n+\theta)} - 5$$

$$\leq \sum_{i=1}^{n} E[|\xi_i - p_i|^3]$$

$$\leq \theta\left\{\log\left(1+\frac{n}{\theta}\right) - \frac{5}{3} + \frac{3\theta}{n+\theta} - \frac{2\theta^2}{(n+\theta)^2} + \frac{2\theta^3}{3(n+\theta)^3}\right\} + 4 + \frac{n}{n+\theta}.$$

(ii) If $n^2/\theta \to \infty$, then it holds that

$$\sum_{i=1}^{n} E[|\xi_i - p_i|^3] \sim \theta\left\{\log\left(1+\frac{n}{\theta}\right) - \frac{5}{3} + \frac{3\theta}{n+\theta} - \frac{2\theta^2}{(n+\theta)^2} + \frac{2\theta^3}{3(n+\theta)^3}\right\}.$$

(iii) In particular, it holds that

$$\sum_{i=1}^{n} E[|\xi_i - p_i|^3] \sim \begin{cases} \theta\log\left(\frac{n}{\theta}\right) & \text{(Case A),} \\ \theta\left\{\log(1+c) - \frac{5}{3} + \frac{3}{c+1} - \frac{2}{(c+1)^2} + \frac{2}{3(c+1)^3}\right\} & \text{(Case B),} \\ \frac{n^2}{2\theta} & \text{(Case C).} \end{cases}$$

Proof (i) The desired inequality is an immediate consequence of (3.13) and Lemma 3.5.

(ii) Because

$$\log(1+x) - \frac{5}{3} + \frac{3}{x+1} - \frac{2}{(x+1)^2} + \frac{2}{3(x+1)^3} > 0$$

for any $x > 0$, it holds that

$$\theta \left\{ \log \left(1 + \frac{n}{\theta} \right) - \frac{5}{3} + \frac{3\theta}{n+\theta} - \frac{2\theta^2}{(n+\theta)^2} + \frac{2\theta^3}{3(n+\theta)^3} \right\} \to \infty,$$

whereas the remainder terms in the inequality of (i) does not diverge to $\pm\infty$. This implies assertion (ii).

(iii) Assertion (iii) is a direct consequence of (ii) (for Case C, the result follows from the Taylor expansion of $\log(1+x) - 5/3 + 3/(x+1) - 2/(x+1)^2 + 2/\{3(x+1)^3\}$ as $x \to 0$). $\qquad\square$

Lemma 3.3 *(i) It holds that*

$$\theta \left\{ \log \left(1 + \frac{n}{\theta} \right) - 2 + \frac{3\theta}{n+\theta} - \frac{\theta^2}{(n+\theta)^2} \right\} + \frac{n}{2(n+\theta)} - 3$$

$$\leq \sum_{i=1}^{n} E[(\xi_i - p_i)^3]$$

$$\leq \theta \left\{ \log \left(1 + \frac{n}{\theta} \right) - 2 + \frac{3\theta}{n+\theta} - \frac{\theta^2}{(n+\theta)^2} \right\} + 2 + \frac{n}{n+\theta}.$$

(ii) In Case A, B*, or C, it holds that

$$\sum_{i=1}^{n} E[(\xi_i - p_i)^3] \sim \begin{cases} \theta \log \left(\frac{n}{\theta} \right) & \text{(Case A),} \\ \theta \left\{ \log(1+c) - 2 + \frac{3}{c+1} - \frac{1}{(c+1)^2} \right\} & \text{(Case B*),} \\ -\frac{n^2}{2\theta} & \text{(Case C).} \end{cases}$$

Proof (i) The desired inequality is an immediate consequence of (3.14) and Lemma 3.5.

(ii) In Case A, the assertion holds because

$$\theta \left\{ \log \left(1 + \frac{n}{\theta} \right) - 2 + \frac{3\theta}{n+\theta} - \frac{\theta^2}{(n+\theta)^2} \right\} \sim \theta \log \left(1 + \frac{n}{\theta} \right) \to \infty,$$

whereas the remainder terms in the inequality of (i) does not diverge to $\pm\infty$. In Case B*, the assertion holds because

$$\log \left(1 + \frac{n}{\theta} \right) - 2 + \frac{3\theta}{n+\theta} - \frac{\theta^2}{(n+\theta)^2} \to \log(1+c) - 2 + \frac{3}{c+1} - \frac{1}{(c+1)^2} \neq 0$$

and $\theta \to \infty$, whereas the remainder terms in the inequality of (i) does not diverge to $\pm\infty$. In Case C, the assertion holds because

$$\theta \left\{ \log \left(1 + \frac{n}{\theta} \right) - 2 + \frac{3\theta}{n+\theta} - \frac{\theta^2}{(n+\theta)^2} \right\} \sim \frac{-n^2}{2\theta} \to -\infty,$$

whereas the remainder terms in the inequality of (i) do not diverge to $\pm\infty$. □

Lemma 3.4 *It holds that*

$$\sum_{i=1}^{n} \left(E[|\xi_i - p_i|^2] \right)^2 \le \theta \left\{ \frac{1}{3} - \frac{\theta}{n+\theta} + \frac{\theta^2}{(n+\theta)^2} - \frac{\theta^3}{3(n+\theta)^3} \right\} + 2. \quad (3.16)$$

Proof The assertion is an immediate consequence of (3.15) and Lemma 3.5. □

Remark 3.9 The asymptotic value of the RHS in (3.16) is given by $\theta/3$ (Case A), $\theta \left[1/3 - 1/(c+1) + 1/(c+1)^2 - 1/\{3(c+1)^3\} \right]$ (Case B), or $n^3/(3\theta^2) + 2$ (Case C).

3.4 Proofs of the Results in Sect. 3.2

3.4.1 Proof of the Results in Sect. 3.2.1

In this subsection, we provide proofs of the results in Sect. 3.2.1.

Proof (Theorem 3.1) Let n be an arbitrary integer such that $n \ge n_0$. From (3.11), Lemma 3.7 yields that

$$\| F_{n,\theta} - \Phi \|_\infty \le C \frac{\sum_{i=1}^{n} E[|\xi_i - p_i|^3]}{(\sum_{i=1}^{n} E[|\xi_i - p_i|^2])^{3/2}},$$

where C is the constant appearing in Lemma 3.7. Additionally, Lemmas 3.1-(i) and 3.2-(i) yield that

$$\frac{\sum_{i=1}^{n} E[|\xi_i - p_i|^3]}{(\sum_{i=1}^{n} E[|\xi_i - p_i|^2])^{3/2}} \le \gamma_1.$$

□

Proof (Corollary 3.1) In Case A, B, or C1, it holds that

$$\theta \left\{ \log \left(1 + \frac{n}{\theta} \right) - \frac{5}{3} + \frac{3\theta}{n+\theta} - \frac{2\theta^2}{(n+\theta)^2} + \frac{2\theta^3}{3(n+\theta)^3} \right\}$$
$$\asymp \theta \left(\log \left(1 + \frac{n}{\theta} \right) - 1 + \frac{\theta}{n+\theta} \right).$$

Hence, Theorem 3.1 and Lemmas 3.1 and 3.2 yield that

$$\gamma_1 = O\left(\frac{1}{\sqrt{\theta\{\log(1 + n/\theta) - 1 + \theta/(n+\theta)\}}}\right).$$

□

Proof (Corollary 3.2) From

$$G_{n,\theta}(x) = F_{n,\theta}\left(\frac{\sigma_T}{\sigma_0}x + \frac{\mu_T - \mu_0}{\sigma_0}\right) \quad (x \in \mathbb{R})$$

and the triangle inequality, it follows that

$$
\begin{aligned}
\|G_{n,\theta} - \Phi\|_\infty &= \sup_{x\in\mathbb{R}}\left|F_{n,\theta}\left(\frac{\sigma_T}{\sigma_0}x + \frac{\mu_T - \mu_0}{\sigma_0}\right) - \Phi(x)\right| \\
&\leq \sup_{x\in\mathbb{R}}\left|F_{n,\theta}\left(\frac{\sigma_T}{\sigma_0}x + \frac{\mu_T - \mu_0}{\sigma_0}\right) - \Phi\left(\frac{\sigma_T}{\sigma_0}x + \frac{\mu_T - \mu_0}{\sigma_0}\right)\right| \\
&\quad + \sup_{x\in\mathbb{R}}\left|\Phi\left(\frac{\sigma_T}{\sigma_0}x + \frac{\mu_T - \mu_0}{\sigma_0}\right) - \Phi\left(\frac{\sigma_T}{\sigma_0}x\right)\right| + \sup_{x\in\mathbb{R}}\left|\Phi\left(\frac{\sigma_T}{\sigma_0}x\right) - \Phi(x)\right|.
\end{aligned}
$$

$$(3.17)$$

The first term on the RHS in (3.17) is

$$O\left(\frac{1}{\sqrt{\theta\{\log(1 + n/\theta) - 1 + \theta/(n+\theta)\}}}\right)$$

from Corollary 3.1. The second term on the RHS in (3.17) is bounded above by

$$\frac{1}{\sqrt{2\pi}}\frac{|\mu_T - \mu_0|}{\sigma_0} = O\left(\frac{1}{\sqrt{\theta\{\log(1 + n/\theta) - 1 + \theta/(n+\theta)\}}}\right)$$

from Lemma 3.6-(i). This is because $|\mu_T - \mu_0| = O(1)$ (Lemma 3.5) and $\sigma_0^2 \sim \theta(\log(1 + n/\theta) - 1 + \theta/(n+\theta))$ (Lemma 3.1). The third term of the RHS in (3.17) is bounded above by

$$\frac{1}{\sqrt{2\pi e}}\max\left(\frac{\sigma_T}{\sigma_0}, 1\right)\left|1 - \frac{\sigma_0}{\sigma_T}\right| = O\left(\frac{1}{\theta\{\log(1 + n/\theta) - 1 + \theta/(n+\theta)\}}\right)$$

from Lemma 3.6-(ii). This is because, from $\sigma_0^2 \geq 1 \geq n/(n+\theta)$ for $n \geq n_1$ and

$$\sigma_0^2 - \frac{n}{\theta + n} \leq \sigma_T^2 \leq \sigma_0^2 - \frac{n}{2(\theta + n)} + 1$$

(see Lemma 3.1-(i)), it follows that

$$\left(1 - \frac{1}{\sigma_0^2}\frac{n}{\theta+n}\right)^{1/2} \le \frac{\sigma_T}{\sigma_0} \le \left[1 + \frac{1}{\sigma_0^2}\left\{1 - \frac{n}{2(\theta+n)}\right\}\right]^{1/2} \tag{3.18}$$

for $n \ge n_1$. Note that the LHS and RHS of (3.18) are

$$1 + O\left(\frac{1}{\sigma_0^2}\right) = 1 + O\left(\frac{1}{\theta\{\log(1+n/\theta) - 1 + \theta/(n+\theta)\}}\right). \qquad \square$$

3.4.2 Proof of the Results in Sect. 3.2.2

In this subsection, we provide proofs of the results in Sect. 3.2.2.

Proof (Theorem 3.2) (i) Let n be an arbitrary integer such that $n \ge n_1$. Because $|\xi_i - p_i| < 1 \le \mathrm{var}(K)$ for all $i = 1, \ldots, n$, (3.11) and Lemma 3.8 yield that

$$\|F_{n,\theta} - \Phi\|_\infty \ge \frac{1}{D}\frac{\left|\sum_{i=1}^n E[(\xi_i - p_i)^3]\right|}{\left(\sum_{i=1}^n E[|\xi_i - p_i|^2]\right)^{3/2}} - \frac{\sum_{i=1}^n \left(E[(\xi_i - p_i)^2]\right)^2}{\left(\sum_{i=1}^n E[|\xi_i - p_i|^2]\right)^2}, \tag{3.19}$$

where D is the constant appearing in Lemma 3.8. Additionally, Lemmas 3.1-(i) and 3.3-(i) yield that

$$\frac{\sum_{i=1}^n E[(\xi_i - p_i)^3]}{\left(\sum_{i=1}^n E[|\xi_i - p_i|^2]\right)^{3/2}} \ge \gamma_2 > 0.$$

Moreover, Lemmas 3.1-(i) and 3.4-(i) yield that

$$\frac{\sum_{i=1}^n \left(E[(\xi_i - p_i)^2]\right)^2}{\left(\sum_{i=1}^n E[|\xi_i - p_i|^2]\right)^2} \le \gamma_3. \tag{3.20}$$

This completes the proof of (i).

(ii) Let n be an arbitrary integer such that $n \ge n_2$. For the same reason given in (i), (3.11) and Lemma 3.8 yield (3.19). Additionally, Lemmas 3.1-(i) and 3.3-(i) yield that

$$-\frac{\sum_{i=1}^n E[(\xi_i - p_i)^3]}{\left(\sum_{i=1}^n E[|\xi_i - p_i|^2]\right)^{3/2}} \ge \gamma_4 > 0.$$

Moreover, Lemmas 3.1-(i) and 3.4-(i) yield (3.20). This completes the proof of (ii).

Proof (Corollary 3.3) In Case A, it follows from

$$\theta \left\{ \log \left(1 + \frac{n}{\theta}\right) - 2 + \frac{3\theta}{n+\theta} - \frac{\theta^2}{(n+\theta)^2} \right\} \sim \theta \left(\log \left(1 + \frac{n}{\theta}\right) - 1 + \frac{\theta}{n+\theta} \right) \sim \theta \log \left(\frac{n}{\theta}\right)$$

that

$$\gamma_2 = O \left(\frac{1}{\sqrt{\theta \log(n/\theta)}} \right).$$

Moreover, it holds that

$$\gamma_3 = O \left(\frac{1}{\theta (\log(1 + n/\theta))^2} \right).$$

Hence, Corollary 3.1 and Theorem 3.2 yield the desired result in Case A.

In Case B⋆, it follows from

$$\theta \left| \log \left(1 + \frac{n}{\theta}\right) - 2 + \frac{3\theta}{n+\theta} - \frac{\theta^2}{(n+\theta)^2} \right| \asymp \theta \left(\log \left(1 + \frac{n}{\theta}\right) - 1 + \frac{\theta}{n+\theta} \right) \asymp \theta$$

that

$$\gamma_2 \sim \gamma_4 = O \left(\frac{1}{\sqrt{\theta}} \right).$$

Moreover, it holds that

$$\gamma_3 = O \left(\frac{1}{\theta} \right).$$

Because

$$\theta \left| \log \left(1 + \frac{n}{\theta}\right) - 2 + \frac{3\theta}{n+\theta} - \frac{\theta^2}{(n+\theta)^2} \right| \to \infty,$$

either n_2 or n_3 exist in Case B⋆. Hence, Corollary 3.1 and Theorem 3.2 yield the desired result in Case B⋆.

In Case C1, it follows from

$$-\theta \left\{ \log \left(1 + \frac{n}{\theta}\right) - 2 + \frac{3\theta}{n+\theta} - \frac{\theta^2}{(n+\theta)^2} \right\} \sim \theta \left(\log \left(1 + \frac{n}{\theta}\right) - 1 + \frac{\theta}{n+\theta} \right) \sim \frac{n^2}{2\theta}$$

that

$$\gamma_4 = O \left(\frac{1}{\sqrt{n^2/\theta}} \right).$$

Moreover, it holds that

$$\gamma_3 \sim \frac{n^3/(3\theta^2) + 2}{n^4/(4\theta^2)} = O\left(\frac{1}{n} + \frac{1}{n^4/\theta^2}\right).$$

Hence, Corollary 3.1 and Theorem 3.2 yield the desired result in Case C1.

3.5 Conclusion

In this paper, we evaluated the approximation errors $\|F_{n,\theta} - \Phi\|_\infty$ and $\|G_{n,\theta} - \Phi\|_\infty$. Deriving decay rates for $\|F_{n,\theta} - \Phi\|_\infty$ when $n/\theta \to c^\star$ (i.e., Case B with $c = c^\star$) and for $\|G_{n,\theta} - \Phi\|_\infty$ is left for future research. Moreover, because normal approximations are refined using the Edgeworth expansion, it is important to derive the Edgeworth expansion under our asymptotic regimes.

Acknowledgements This work was partly supported by the Japan Society for the Promotion of Science KAKENHI Grant Number 16H02791, 18K13454.

Appendix 1. Some Evaluations

The following lemma is used in the main body.

Lemma 3.5 *Let θ be a positive value and n be a positive integer. (i) It holds that*

$$\log\left(1 + \frac{n}{\theta}\right) + \frac{n}{2\theta(n+\theta)} \le \sum_{i=1}^{n} \frac{1}{\theta + i - 1} \le \log\left(1 + \frac{n}{\theta}\right) + \frac{n}{\theta(n+\theta)}.$$

(ii) It holds that

$$\frac{1}{k\theta^k} - \frac{1}{k(n+\theta)^k} \le \sum_{i=1}^{n} \frac{1}{(\theta + i - 1)^{k+1}} \le \frac{1}{\theta^{k+1}} + \frac{1}{k\theta^k} - \frac{1}{k(n+\theta)^k}$$

for any positive integer k.

Proof For (i), see Tsukuda [15, Proof of Proposition 1]. For (ii), the conclusion follows from

$$\int_\theta^{n+\theta} \frac{dx}{x^{k+1}} \le \sum_{i=1}^{n} \frac{1}{(\theta + i - 1)^{k+1}} \le \frac{1}{\theta^{k+1}} + \int_\theta^{n+\theta} \frac{dx}{x^{k+1}}$$

for any positive integer k. $\qquad\square$

The next lemma provides basic results on the standard normal distribution function.

Lemma 3.6 *(i) For any $\alpha(\in \mathbb{R})$, it holds that*

$$\sup_{x\in\mathbb{R}} |\Phi(x+\alpha) - \Phi(x)| \le \frac{|\alpha|}{\sqrt{2\pi}}.$$

(ii) For any positive $\beta(\in \mathbb{R})$, it holds that

$$\sup_{x\in\mathbb{R}} |\Phi(\beta x) - \Phi(x)| \le \max(\beta, 1)\frac{|1 - 1/\beta|}{\sqrt{2\pi e}}.$$

Proof (i) For some δ between 0 and α, it holds that

$$|\Phi(x+\alpha) - \Phi(x)| = \phi(x+\delta)|\alpha| \le \sup_{x\in\mathbb{R}} \phi(x)|\alpha| = \frac{|\alpha|}{\sqrt{2\pi}}.$$

(ii) Because

$$\max(\beta, 1)\frac{|1 - 1/\beta|}{\sqrt{2\pi e}} = \begin{cases} (\beta - 1)/\sqrt{2\pi e} & (\beta \ge 1) \\ (1/\beta - 1)/\sqrt{2\pi e} & (0 < \beta \le 1), \end{cases}$$

we prove the assertions for $\beta \ge 1$ and $0 < \beta \le 1$ separately. First, we consider the case $\beta \ge 1$. For $x = 0$, it holds that $|\Phi(\beta x) - \Phi(x)| = 0$. For $x > 0$,

$$|\Phi(\beta x) - \Phi(x)| = \int_x^{\beta x} \phi(t)dt$$
$$\le \phi(x)(\beta - 1)x = -\phi'(x)(\beta - 1) \le \sup_{x>0}(-\phi'(x))(\beta - 1) = \frac{\beta - 1}{\sqrt{2\pi e}}.$$

For $x < 0$,

$$|\Phi(\beta x) - \Phi(x)| = \int_{\beta x}^{x} \phi(t)dt$$
$$\le \phi(x)(1 - \beta)x = \phi'(x)(\beta - 1) \le \sup_{x<0}(\phi'(x))(\beta - 1) = \frac{\beta - 1}{\sqrt{2\pi e}}.$$

Next, we consider the case $0 < \beta \le 1$. For $x = 0$, it holds that $|\Phi(\beta x) - \Phi(x)| = 0$. For $x > 0$,

$$|\Phi(\beta x) - \Phi(x)| = \int_{\beta x}^{x} \phi(t)dt$$
$$\le \phi(\beta x)(1 - \beta)x = -\phi'(\beta x)\left(\frac{1}{\beta} - 1\right) \le \sup_{x>0}(-\phi'(x))\left(\frac{1}{\beta} - 1\right) = \frac{1/\beta - 1}{\sqrt{2\pi e}}.$$

For $x < 0$,

$$|\Phi(\beta x) - \Phi(x)| = \int_x^{\beta x} \phi(t)dt$$

$$\leq \phi(\beta x)(\beta - 1)x = \phi'(\beta x)\left(\frac{1}{\beta} - 1\right) \leq \sup_{x<0}(\phi'(x))\left(\frac{1}{\beta} - 1\right) = \frac{1/\beta - 1}{\sqrt{2\pi e}}. \qquad \square$$

Appendix 2. Error Bounds for Normal Approximations

The Berry–Esseen-Type Theorem for Independent Sequences

In this subsection, we introduce the Berry–Esseen-type theorem for independent sequences. For further details, see Tyurin [17].

Let $\{X_i\}_{i\geq 1}$ be a sequence of independent random variables, where $E[X_i] = 0$, $E[X_i^2] = \sigma_i^2 (> 0)$, and $E[|X_i|^3] = \beta_i < \infty$, for all $i = 1, 2, \ldots$ The quantity $\varepsilon_n = \sum_{i=1}^n \beta_i / (\sum_{i=1}^n \sigma_i^2)^{3/2}$ is called the Lyapunov fraction. We denote the distribution function of $\sum_{i=1}^n X_i / (\sum_{i=1}^n \sigma_i^2)^{1/2}$ by F_n^X. Then, the following result holds.

Lemma 3.7 (Tyurin [17]) *There exists a universal constant C such that*

$$\|F_n^X - \Phi\|_\infty \leq C\varepsilon_n$$

for all positive integers n, where C does not exceed 0.5591.

Remark 3.10 Here, we introduce the result given by Tyurin [17]. Many studies have derived Berry–Esseen-type results; see, for example, Chen et al. [2, Chap. 3].

Lower Bound

In this subsection, we introduce the result given by Hall and Barbour [9] that considers reversing the Berry–Esseen inequality.

Let $\{Y_i\}_{i\geq 1}$ be a sequence of independent random variables satisfying $E[Y_i] = 0$ and $E[Y_i^2] = \sigma_i^2 (> 0)$ for all i, and $\sum_{i=1}^n \sigma_i^2 = 1$. We denote the distribution function of $\sum_{i=1}^n Y_i$ by F_n^Y. Letting

$$\delta = \sum_{i=1}^n E[Y_i^2 \mathbb{I}\{|Y_i| > 1\}] + \sum_{i=1}^n E[Y_i^4 \mathbb{I}\{|Y_i| \leq 1\}] + \left|\sum_{i=1}^n E[Y_i^3 \mathbb{I}\{|Y_i| \leq 1\}]\right|,$$

the following result holds.

Lemma 3.8 (Hall and Barbour [9]) *There exists a universal constant D such that*

$$\delta \le D\left(\|F_n^Y - \Phi\|_\infty + \sum_{i=1}^n \sigma_i^4 \right).$$

Because

$$\delta \ge \sum_{i=1}^n \mathrm{E}[Y_i^2 \mathbb{I}\{|Y_i| > 1\}] + \left| \sum_{i=1}^n \mathrm{E}[Y_i^3 \mathbb{I}\{|Y_i| \le 1\}] \right|,$$

we use the RHS as a lower bound. This bound is sufficient in Cases A, B*, and C1 to show the decay rate of $\|F_{n,\theta} - \Phi\|_\infty$.

References

1. Arratia R, Tavaré S (1992) Limit theorems for combinatorial structures via discrete process approximations. Random Struct Algor 3(3):321–345
2. Chen LHY, Goldstein L, Shao Q-M (2011) Normal approximation by Stein's method. Springer, Heidelberg
3. Crane H (2016) The ubiquitous Ewens sampling formula. Statist Sci 31(1):1–19
4. Durrett R (2008) Probability models for DNA sequence evolution, 2nd edn. Springer, New York
5. Ewens WJ (1972) The sampling theory of selectively neutral alleles. Theoret Population Biology 3:87–112; erratum, ibid 3: 240 (1972); erratum, ibid 3: 376 (1972)
6. Feng S (2007) Large deviations associated with Poisson-Dirichlet distribution and Ewens sampling formula. Ann Appl Probab 17(5–6):1570–1595
7. Ghosal S, van der Vaart A (2017) Fundamentals of nonparametric Bayesian inference. Cambridge University Press, Cambridge
8. Goncharov VL (1944) Some facts from combinatorics. Izv Akad Nauk SSSR Ser Mat 8:3–48
9. Hall P, Barbour AD (1984) Reversing the Berry-Esseen inequality. Proc Amer Math Soc 90(1):107–110
10. Hwang H-K (1998) On convergence rates in the central limit theorems for combinatorial structures. European J Combin 19(3):329–343
11. Johnson NL, Kotz S, Balakrishnan N (1997) Discrete multivariate distributions. Wiley, New York
12. Kabluchko Z, Marynych A, Sulzbach H (2016) Mode and Edgeworth expansion for the Ewens distribution and the Stirling numbers. J Integer Seq 19(8): Art. 16.8.8, 17 pp
13. Sibuya M (1986) Stirling family of probability distributions. JPN J Appl Stat 15(3):131–146 (in Japanese)
14. Sibuya M (1988) Log-concavity of Stirling numbers and unimodality of Stirling distributions. Ann Inst Statist Math 40(4):693–714
15. Tsukuda K (2017) Estimating the large mutation parameter of the Ewens sampling formula. J Appl Probab 54(1):42–54; correction, ibid 55(3):998–999 (2018)
16. Tsukuda K (2019) On Poisson approximations for the Ewens sampling formula when the mutation parameter grows with the sample size. Ann Appl Probab 29(2):1188–1232
17. Tyurin IS (2012) A refinement of the remainder in the Lyapunov theorem. Theory Probab Appl 56(4):693–696
18. Watterson GA (1974) Models for the logarithmic species abundance distributions. Theoret Popul Biol 6:217–250

19. Watterson GA (1974) The sampling theory of selectively neutral alleles. Adv Appl Probab
    6:463–488
20. Yamato H (2013) Edgeworth expansions for the number of distinct components associated with
    the Ewens sampling formula. J Japan Statist Soc 43(1):17–28
21. Yamato H, Nomachi T, Toda K (2015) Approximate distributions for the number of dis-
    tinct components of the Ewens sampling formula and its applications. Bull Inform Cybernet
    47:69–81

Chapter 4
Distribution of Number of Levels in an [s]-Specified Random Permutation

James C. Fu

Abstract Successions, Eulerian and Simon Newcomb numbers, and levels are the best-known patterns associated with [s]-specified random permutations. The distribution of the number of rises was first studied in 1755 by Euler. However, the distribution of the number of levels in an [s]-specified random permutation remained unknown. In this study, our main goal is to identify the distribution of the number of levels, which we achieve using the finite Markov chain imbedding technique and insertion procedure. An example is given to illustrate the theoretical result.

Keywords Eulerian and Newcomb numbers · Finite Markov chain imbedding · Insertion · Levels · Random permutation · [s]-specified random permutation

4.1 Introduction

Let $S_N = \{1, \ldots, 1, 2, \ldots, 2, \ldots, n, \ldots, n\}$ be a set of N integers with specification $[s] = (s_1, \ldots, s_n)$, where $s_i (\geq 1)$ is the number of integers i, for $i = 1, \ldots, n$, in set S_N, and $s_1 + \cdots + s_n = N$. Let $\mathcal{H}(N) = \{\boldsymbol{\pi}(N) = (\pi_1(N), \ldots, \pi_N(N)) : \pi_t(N) \in S_N, t = 1, \ldots, N\}$ be the collection of all permutations generated by the integers in S_N with an [s]-specification: for example, $S_5 = \{1, 1, 2, 2, 3\}$, $[s] = [2, 2, 1]$, and $\mathcal{H}(5) = \{\boldsymbol{\pi}(5) = (\pi_1(5), \ldots, \pi_5(5)) = (1, 2, 3, 1, 2) : \pi_i(5) \in S_5, i = 1, \ldots, 5\}$. For $t = 1, \ldots, N-1$, we define the rises (increases) $R_{t,N}$, falls (decreases) $D_{t,N}$, and levels $L_{t,N}$ for the gaps among $\pi_1(N)$ to $\pi_N(N)$ as three indicator functions:

J. C. Fu (✉)
Department of Statistics, University of Manitoba, Winnipeg R3T 2N2, Canada
e-mail: James.Fu@umanitoba.ca

© The Author(s), under exclusive license to Springer Nature Singapore Pte Ltd. 2020   75
N. Hoshino et al. (eds.), *Pioneering Works on Distribution Theory*,
JSS Research Series in Statistics,
https://doi.org/10.1007/978-981-15-9663-6_4

$$R_{t,N} = \begin{cases} 1 & \text{if } \pi_{t+1}(N) > \pi_t(N) \\ 0 & \text{otherwise,} \end{cases}$$

$$D_{t,N} = \begin{cases} 1 & \text{if } \pi_{t+1}(N) < \pi_t(N) \\ 0 & \text{otherwise, and} \end{cases} \qquad (4.1)$$

$$L_{t,N} = \begin{cases} 1 & \text{if } \pi_{t+1}(N) = \pi_t(N) \\ 0 & \text{otherwise.} \end{cases}$$

By convention, we let the front end be a rise and the rear end be a fall; that is, $R_{0,N} \equiv 1$ and $D_{N,N} \equiv 1$. Furthermore, we define $R_N = \sum_{t=0}^{N-1} R_{t,N}$, $D_N = \sum_{t=1}^{N} D_{t,N}$, and $L_N = \sum_{t=1}^{N-1} L_{t,N}$ as the total number of rises, falls, and levels, respectively, associated with random permutations $\pi(N) \in \mathcal{H}(N)$.

The numbers of patterns (e.g., successions, rises, and falls) associated with random permutations are used successfully in areas such as statistics, applied probability, and discrete mathematics, although their theoretical development began in the eighteenth century; see, for example, [7]. Their history extensions, and applications can be found in the books on combinatorial analysis by [4, 5, 16, 18]. Despite their long history, the numbers of successions, rises, and falls (e.g., [1–3, 7, 8, 10, 11, 13–15, 17, 19, 20]) remain of considerable interest to researchers.

Let $A([s], k)$ be the number of permutations in $\mathcal{H}(N)$ having exactly k rises. The number $A([s], k)$ is referred to as the Simon Newcomb number. For the special case of $s_i \equiv 1$, for $i = 1, \ldots, n$, the Simon Newcomb number is referred to as Eulerian number $A(n, k)$, developed by Euler in his famous book "*Institutiones calculi differentialis*" (1755, pp. 485–487), and is defined as the solution of the following identity:

$$\frac{1 - x}{1 - x \exp\{\lambda(1 - x)\}} = \sum_{n=0}^{\infty} \sum_{k=1}^{\infty} A(n, k) x^k \frac{\lambda^n}{n!}. \qquad (4.2)$$

The formula for $A(n, k)$ has the following form:

$$A(n, k) = \sum_{j=0}^{k} (-1)^j (k - j)^n \binom{n + 1}{j}, \quad n \geq k \geq 1, \qquad (4.3)$$

which also satisfies the recursive equation

$$A(n, k) = k A(n - 1, k) + (n - k + 1) A(n - 1, k - 1). \qquad (4.4)$$

Dillon and Roselle (1969) [6] gave the following formula for the Simon Newcomb numbers:

$$A([s], k) = \sum_{j=0}^{k} (-1)^j \binom{N + 1}{j} \prod_{i=1}^{n} \binom{s_i + k - j - 1}{s_i}. \qquad (4.5)$$

Fu et al. (1999) [11] used the finite Markov chain imbedding technique [9] and insertion procedure to show that the random variable R_N is nonhomogeneous Markov chain imbeddable and

$$P(R_N = k) = \xi_0 \left(\prod_{t=1}^{N} M_t([s]) \right) U_k',$$ (4.6)

where $\xi_0 = (1, 0, \ldots, 0)$, $U_k = (0, \ldots, 0, 1, 0, \ldots, 0)$ is the kth unit vector, and $M_t([s])$, for $t = 1, \ldots, N$, are the transition probability matrices of the imbedded Markov chain. Using a last-step analysis and the result for Eq. (4.6), they obtained the following simple recursive equation:

$$P(R_N = k) = \frac{N - \ell(N) - k + 1}{N} P(R_{N-1} = k - 1) + \frac{k + \ell(N)}{N} P(R_{N-1} = k),$$ (4.7)

where $\ell(N) = N - \sum_{i=1}^{n-1} s_i - 1$. Hence, Eq. (4.7) yields the following recursive equation for the Simon Newcomb numbers:

$$s_n A([s], k) = (N - \ell(N) - k + 1) A([s'], k - 1) + (k + \ell(N)) A([s'], k),$$ (4.8)

where $[s'] = (s_1, \ldots, s_{n-1}, s_n - 1)$. The recursive equation (4.8) is rather simple, but compute the Simon Newcomb numbers very fast. To the best of my knowledge, no recursive equations exist to determine for the Simon Newcomb numbers in combinatorial analysis.

For the special case $[s] = [1, \ldots, 1]$, it follows that $\ell(t) \equiv 0$, for $t = 1, \ldots, n$; thus, the transition probability matrices are given by

$$M_t = \begin{array}{c} \\ 1 \\ \vdots \\ k \\ \vdots \\ t \end{array} \begin{array}{c} \quad 1 \cdots k \cdots t-1\ t \\ \left[\begin{array}{ccccc} \frac{1}{t} & \frac{t-1}{t} & & & \\ & \ddots & \ddots & & 0 \\ & & \frac{k}{t} & \frac{t-k}{t} & \\ 0 & & & \ddots & \ddots \\ & & & & \frac{t-1}{t}\ \frac{1}{t} \end{array} \right] \end{array},$$

for all $t = 1, \ldots, n$, and the Eulerian number is given as

$$A(n, k) = n! \xi_0 \left(\prod_{t=1}^{n} M_t \right) U_k'.$$

Furthermore, because $n = N$ and $s_n = 1$, the recursive equation (4.8) yields Eq. (4.4).

In contrast, determining the distribution of the number of levels in an [s]-specified random permutation is rather complex. To the best of my knowledge, this problem remains unsolved, likely owing to the complexity of the structure of

$[s] = (s_1, \ldots, s_n)$, and because the number of levels is independent of the order of integers. It seems that using a combinatorial method to obtain the exact formulae for the exact distribution of levels is tedious, if not impossible. The mathematical challenge of solving this problem is the main motivation of this study. Apart from the combinatorial approach, we adopt the finite Markov chain imbedding technique and insertion procedure to construct a nonhomogeneous Markov chain $\{Y_t\}_1^N$, which we use to obtain the distribution of the number of levels. To do so, we determine the probability of projecting the imbedded Markov chain onto a subspace of the state space c_ℓ; that is,

$$P(L_N = \ell) = P(Y_N \in c_\ell). \tag{4.9}$$

A detailed numerical example is provided to illustrate the theoretical results.

4.2   Distribution of Number of Levels

Let $\mathcal{H}'(N)$ be the collection of $(s_1', \ldots, s_n')$-specified random permutations with $s_{\ell_1}' \geq \cdots \geq s_{\ell_n}'$ and $s_1' + \cdots + s_n' = N$, and let $\mathcal{H}(N)$ be the collection of all $[s] = (s_1, \ldots, s_n)$-specified random permutations with $s_i = s_{\ell_i}'$, for $i = 1, \ldots, n$. It follows from the one-to-one transformation of index $i \Leftrightarrow \ell_i$, for $i = 1, \ldots, n$, that the distributions of the number of levels of the two random variables L_N' (the number of levels in $\mathcal{H}'(N)$) and L_N are the same. This is because the number of levels in the random permutation is independent of the order among the integers (symbols). Hence, throughout this manuscript, we need only to study the distribution of the number of levels L_N, with $s_1 \geq \cdots \geq s_n \geq 1$.

For given t, $t = 1, \ldots, N$, there exists an integer i, for $i = 1, \ldots, n$, such that

$$t = s_0 + s_1 + \cdots + s_{i-1} + t_i, \tag{4.10}$$

where $1 \leq t_i \leq s_i$ and $s_0 = 0$. Let $\mathcal{H}(t)$ be all random permutations $\pi(t) = (\pi_1(t), \ldots, \pi_t(t))$ generated by the first t integers. For every $\pi(t) \in \mathcal{H}(t)$, this can be viewed as the result of inserting $1, \ldots, 1, \ldots, i, \ldots, i$, one by one, into the gaps between two integers and between the two end gaps. Furthermore, the random permutation $\pi(N)$ is the result of a sequence of realizations $\{\pi(t)\}_{t=1}^N$ of inserting the integers one at a time. For given t, hence t_i, the next integer k ($k = i$ or $k = i + 1$) to be inserted into the $t + 1$ gaps of the random permutation $\pi(t)$ is integer i if $1 \leq t_i < s_i$ and is integer $i + 1$ if $t_i = s_i$.

The procedure for inserting integer k is as follows:

(A) if an integer k is inserted into a gap $[j, j]$ (a gap indicates that two adjacent symbols are j), for $j \neq k$, then the level $[j, j]$ is destroyed,

(B) if an integer k is inserted into a gap such as $[k, k]$, $[k, i]$, $[i, k](i \neq k)$, $[\cdot, k]$, or $[k, \cdot]$, then a new level $[k, k]$ is created, and

(C) if an integer k is inserted into $[i, j]$, for $i \neq j \neq k$, then no level is created or destroyed.

Given $t = s_0 + s_1 + \cdots + s_{i-1} + t_i$ and $\pi(t) \in \mathcal{H}(t)$, let the set $G(\pi(t)) = \{[i, j] : \text{gaps of } \pi(t)\}$ be $t + 1$ gaps of random permutations of $\pi(t)$, including two end gaps. By convention, we assume that the two end gaps are $[\cdot, j]$ and $[h, \cdot]$, for $1 \leq j$ and $h \leq n$, throughout this manuscript. Note that the integer k can only be i if $1 \leq t_i < s_i$; otherwise, $k = i + 1$ if $t_i = s_i$. For $k = i$, it follows from rules (A), (B), and (C) that the set of $t + 1$ gaps $G(\pi(t))$ can be partitioned into three subsets:

$$D_t(i, i) = \{\text{all gaps in } G(\pi(t)) \text{ and has the form } [j, j], \text{ with } j < i\},$$
$$I_t(i, i) = \{\text{all gaps in } G(\pi(t)) \text{ and has the form } [i, i], [i, j], \text{ or}$$
$$[j, i], \text{ with } j < i\},$$
$$N_t(i, i) = \{\text{all gaps in } G(\pi(t)) \text{ and has the form } [h, j], \text{ with } h \neq j \text{ and}$$
$$h, j < i\}, \text{ and}$$
$$G(\pi(t)) = D_t(i, i) \cup I_t(i, i) \cup N_t(i, i). \tag{4.11}$$

Similarly, for $k = i + 1$, the set $G(\pi(t))$ can be partitioned into three subsets:

$$D_t(i, i + 1) = \{\text{all gaps in } G(\pi(t)) \text{ and has the form } [j, j], \text{ with } j < i + 1\},$$
$$I_t(i, i + 1) = \{\text{all gaps in } G(\pi(t)) \text{ and has the form } [i + 1, i + 1]\} = \{\emptyset\},$$
$$N_t(i, i + 1) = \{\text{all gaps in } G(\pi(t)) \text{ and has the form } [h, j], \text{ with}$$
$$h \neq j, h, j < (i + 1)\}, \text{ and}$$
$$G(\pi(t)) = D_t(i, i + 1) \cup I_t(i, i + 1) \cup N_t(i, i + 1). \tag{4.12}$$

The following example clarifies the above subsets:
Given $\pi(7) = (1, 1, 2, 2, 1, 2, 1)$, we have eight gaps (including two end gaps), $G(\pi(7)) = \{[\cdot, 1], [1, 1], [1, 2], [2, 2], [2, 1], [1, 2], [2, 1], [1, \cdot]\}$. With respect to the next integer $k(k = 2 \text{ or } 3)$ to be inserted, the above definitions imply the following subsets:

1. for $k = 2, D_7(2, 2) = \{[1, 1]\}, I_7(2, 2) = \{[1, 2], [2, 2], [2, 1], [1, 2], [2, 1]\}$, and $N_7(2, 2) = \{[\cdot, 1], [1, \cdot]\}$, and
2. for $k = 3, D_7(2, 3) = \{[1, 1], [2, 2]\}, I_7(2, 3) = \{\emptyset\}$, and $N_7(2, 3) = \{[\cdot, 1], [1, 2], [2, 1], [1, 2], [2, 1], [1, \cdot]\}$.

Given $k = i$ or $i + 1$, let $\#D_t(i, k), \#I_t(i, k)$, and $\#N_t(i, k)$ be the number of the gaps in the subsets $D_t(i, k), I_t(i, k)$, and $N_t(i, k)$, respectively. Then, the following lemma holds:

Lemma 4.1 *Given $t = s_1 + \cdots + s_{i-1} + t_i$, and the random permutation $\pi(t)$, we have*

$$\#D_t(i, k) = \begin{cases} L_t - L_t(i), & \text{if } k = i, \\ L_t, & \text{if } k = i + 1, \end{cases}$$

$$\#I_t(i, k) = \begin{cases} 2t_i - L_t(i), & \text{if } k = i, \\ 0, & \text{if } k = i + 1, \end{cases}$$

and

$$\#N_t(i, k) = (t + 1) - \#D_t(i, k) - \#I_t(i, k),$$

where L_t is the total number of levels in $\pi(t)$, and $L_t(i)$ is the total number of levels $[i, i]$ in $\pi(t)$.

Proof Given $\pi(t)$, L_t, and $L_t(i)$, the above results follow directly from the integer k to be inserted and the definitions of $D_t(i, k)$, $I_t(i, k)$, and $N_t(i, k)$, and $\#G(\pi(t)) = \#D_t(i, k) + \#I_t(i, k) + \#N_t(i, k) = t + 1$. □

Lemma 4.2 *Given a random permutation $\pi(t) \in \mathcal{H}(t)$ and levels $(L_t, L_t(i))$, we have the following:*

1. *If the integer $k(k = i$ or $i + 1)$ is inserted into one of the gaps of $D_t(i, k)$ $(\langle (L_t, L_t(i)) : k \to D_t(i, k) \rangle)$, then $\pi(t + 1)$ has levels*

$$(L_{t+1}, L_{t+1}(k)) = \langle (L_t, L_t(i)), k \to D_t(i, k) \rangle$$
$$= \begin{cases} (L_t - 1, L_t(i)), & \text{for } k = i, \\ (L_t - 1, 0), & \text{for } k = i + 1; \end{cases}$$

2.

$$(L_{t+1}, L_{t+1}(k)) = \langle (L_t, L_t(i)), k \to I_t(i, k) \rangle$$
$$= \begin{cases} (L_t + 1, L_t(i) + 1), & \text{for } k = i, \\ (L_t, 0), & \text{for } k = i + 1; \end{cases}$$

and

3.

$$(L_{t+1}, L_{t+1}(k)) = \langle (L_t, L_t(i)), k \to N_t(i, k) \rangle$$
$$= \begin{cases} (L_t, L_t(i)), & \text{for } k = i, \\ (L_t, 0), & \text{for } k = i + 1. \end{cases}$$

Proof Note that if the integer $k(k = i$ or $k = i + 1)$ is inserted into the gap $[j, j] \in D_t(i, k)$ and $j < i$, then from rule (A) and the definitions of L_{t+1} and $L_{t+1}(k)$, it follows that we have $(L_{t+1}, L_{t+1}(k)) = (L_t - 1, L_t(i))$ for $k = i$, and $(L_{t+1}, L_{t+1}(k)) = (L_t - 1, 0)$ for $k = i + 1$. This proves part (i). Similarly, results (ii) and (iii) are direct consequence of rules (B) and (C) and the definitions of L_{t+1} and $L_{t+1}(k)$. □

For every $\pi(t) \in \mathcal{H}(t)$, for $t = 1, \ldots, N$, we define

$$Y_t = (L_t, L_t(i)) \tag{4.13}$$

and define the transformation $Y_t \overset{k}{\to} Y_{t+1}$ induced by the insertion procedure (Lemma 4.2), regardless of whether $k = i$ or $i + 1$, as

$$Y_{t+1} = (L_{t+1}, L_{t+1}(k)) = < (L_t, L_t(i)), k \to G(\pi(t)) >, \tag{4.14}$$

where $< \cdot, \cdot >$ obeys Lemma 4.2. Furthermore, given the state space $\Omega_1 = \{(0, 0)\}$, we define the state spaces Ω_t sequentially as

$$\Omega_{t+1} =$$
$$\bigcup_{(L_t, L_t(i)) \in \Omega_t} \{< (L_t, L_t(i)), k \to H(t) >: H_t = D_t(i, k), I_t(i, k) \text{ and } N_t(i, k)\},$$
$$\tag{4.15}$$

for $t = 2, \ldots, N - 1$.

Note that the construction of the state space $\{\Omega_t\}$ Eq. (4.15) depends heavily on the structure of $[s] = (s_1, \ldots, s_n)$.

Lemma 4.3 *If the probabilities of inserting the integer k into $t + 1$ gaps are equally likely, then the transition probabilities are*

$$(L_{t+1}, L_{t+1}(k)) = \begin{cases} < (L_t, L_t(i)), k \to D_t(i, k) >, & \text{with prob. } \frac{\#D_t(i,k)}{t+1}, \\ < (L_t, L_t(i)), k \to I_t(i, k) >, & \text{with prob. } \frac{\#I_t(i,k)}{t+1}, \\ < (L_t, L_t(i)), k \to N_t(i, k) >, & \text{with prob. } \frac{\#N_t(i,k)}{t+1}, \end{cases} \tag{4.16}$$

for $t = 1, \ldots, N - 1$, where $< \cdot, \cdot >$ obeys Lemma 4.2, and $\#D_t(i, k), \#I_t(i, k)$, and $\#N_t(i, k)$ are given by Lemma 4.1.

Proof Given t and k, the above results follow directly from the definition of Y_t, Eqs. (4.14) and (4.15), and Lemmas 4.1 and 4.2. □

Note that the sequence $\{Y_t\}$ defined on $\{\Omega_t\}$ has the following transition probability matrices, defined by Eq. (4.16), for $t = 1, \ldots, N - 1$:

$$\Omega_{t+1}$$
$$M_t = \Omega_t \begin{bmatrix} P(Y_{t+1} = (L_{t+1}, L_{t+1}(k))|Y_t = (L_t, L_t(i))) \end{bmatrix}. \tag{4.17}$$

For convenience, we assume Y_0 at dummy state $\emptyset$ ($\Omega_0 = \{\emptyset\}$), and the state $\emptyset$ goes to state $(0, 0)$ with probability one (transition probability matrix [1]). It follows from Eqs. (4.14)–(4.17) that $\{Y_t\}_0^N$ forms a nonhomogeneous finite Markov chain, with transition probability matrices $\{M_t\}_0^{N-1}$ given by Eq. (4.17). Hence, the random variable L_N is nonhomogeneous finite Markov chain imbeddable, and its distribution is given by the following Theorem:

Theorem 4.1 *Given an* $[s]$*-specified random permutation* $\pi(N)$*, the random variable* L_N *is nonhomogeneous finite Markov chain imbeddable in the sense that there exists a Markov chain* $\{Y_t\}$ *defined on the state space* $\{\Omega_t\}$ *with transition probability matrices* M_t*, defined by Eqs. (4.15), (4.16), and (4.17) and*

$$P(L_N = \ell) = P\left(Y_N \in \{(\ell, 0), \ldots, (\ell, s_n - 1)\}\right)$$
$$= \begin{cases} \xi_0 \left(\prod_{t=0}^{N-1} M_t\right) U'(c_\ell), & \text{for } \max(0, 2s_1 - N - 1) \le \ell \le N - n, \\ 0, & \text{otherwise,} \end{cases}$$

(4.18)

where ξ_0 *is the initial distribution* Y_0 *at the dummy state* $\emptyset$ *with probability one,* $c_\ell = \{(\ell, 0), \ldots, (\ell, s_n - 1)\}$ *is a subset of the state space* Ω_N*, and* $U(c_\ell) = (0, \ldots, 0, 1, \ldots, 1, 0, \ldots, 0)$.

Proof Given $\max(0, 2s_1 - N - 1) \le \ell \le N - n$, it follows from Lemmas 4.1 and 4.3 that $\{Y_t\}$ is a nonhomogeneous Markov chain. Hence, the random variable L_N is finite Markov chain imbeddable. The first part of the result in Eq. (4.18) follows directly from the Chapman−Kolmogorov equation (or Fu and Lou (2003) [12], Theorem 2.1).

Note that there is no random permutation $\pi(N)$ in $\mathcal{H}(N)$ in which the number of levels is less than $\max(0, 2s_1 - N - 1)$; hence, $P(L_N = \ell) = 0$, for all $0 \le \ell < \max(0, 2s_1 - N - 1)$.   □

It follows from the above theorem that the total number of random permutations with ℓ levels in $\mathcal{H}(N)$ is

$$L_N([s], \ell) = \begin{cases} \frac{N!}{\prod_{i=1}^{n} s_i!} \xi_0 \left(\prod_{t=0}^{N-1} M_t\right) U'(c_\ell), & \text{for } \max(0, 2s_1 - N - 1) \le \ell \le N - n, \\ 0, & \text{otherwise.} \end{cases}$$

(4.19)

Theorem 4.1 also provides simple expressions for the moments, $E(L_N^k)$, and generating functions. Letting $N^* = N - n$, $N_* = \max(0, 2s - N - 1)$,

$$E(k) = \sum_{\ell=N_*}^{N^*} \ell^k U'(c_\ell), \quad \text{and} \quad \phi(z) = \sum_{\ell=N_*}^{N^*} z^\ell U'(c_\ell),$$

we have

$$E(L_N^k) = \xi_0 \left(\prod_{t=0}^{N-1} M_t\right) E(k) \quad \text{and} \quad \phi_{L_N}(z) = \xi_0 \left(\prod_{t=0}^{N-1} M_t\right) \phi(z).$$

In order to illustrate our main result, we first provide several technique remarks first, followed by a detailed numerical example.

Remark 4.1 The condition $\max(0, 2s_1 - N - 1) \le \ell \le N - n$ is vital to guarantee that the above theorem covers cases such as those of $\mathcal{H}(N)$ with $2s_1 - N - 1 > 0$; for example, $[s] = [8, 2, 2]$ and $P(K_{L_{12}} = \ell) = 0$, for $0 \le \ell < 3$.

Remark 4.2 The one-to-one indexing transformation means that Theorem 4.1 can be applied to categorical [*s*]-specified random permutations. For example, let s_A, s_C, s_G, and s_T, be the numbers of categories for A, C, G, and T, respectively. Let $s_1 = s_A, s_2 = s_C, s_3 = s_G$, and $s_4 = s_T$; then, the distribution of the number of levels can be derived in the same way as in the previous discussion.

Example

The following detailed example illustrates our main result. Consider a $(4, 2, 1)$-specified random permutation $\in \mathcal{H}(7)$. From Lemmas 4.1 and 4.2, it follows that the state spaces can be determined as follows:
$\Omega_0 = \{\emptyset\}$, $\Omega_1 = \{(0, 0)\}$, $\Omega_2 = \{(1, 1)\}$, $\Omega_3 = \{(2, 2)\}$, $\Omega_4 = \{(3, 3)\}$, $\Omega_5 = \{(2, 0), (3, 0)\}$, $\Omega_6 = \{(1, 0), (2, 0), (3, 0), (3, 1), (4, 1)\}$, and $\Omega_7 = \{(0, 0), (1, 0), (2, 0), (3, 0), (4, 0)\}$. It then follows from Eq. (4.16) in Lemma 4.3 that the imbedded Markov chain $\{Y_t\}_0^7$ defined on the above state spaces has the transition probability matrices

$$M_0 = [1], M_1 = [1], M_2 = [1], M_3 = [1], M_4 = [3/5, 2/5],$$

$$M_5 = \begin{bmatrix} 2/6 & 2/6 & 0 & 2/6 & 0 \\ 0 & 3/6 & 1/6 & 0 & 2/6 \end{bmatrix} \text{ and}$$

$$M_6 = \begin{bmatrix} 1/7 & 6/7 & 0 & 0 & 0 \\ 0 & 2/7 & 5/7 & 0 & 0 \\ 0 & 0 & 3/7 & 4/7 & 0 \\ 0 & 0 & 3/7 & 4/7 & 0 \\ 0 & 0 & 0 & 4/7 & 3/7 \end{bmatrix}.$$

It follows from Eqs. (4.18) and (4.19) that we have

ℓ	0	1	2	3	4	Total
$P(L_N = \ell)$	1/35	10/35	14/35	8/35	2/35	1
$L_N([s], \ell)$	3	30	42	24	6	105

For example, there are three random permutations $\{(1, 2, 1, 2, 1, 3, 1), (1, 2, 1, 3, 1, 2, 1), (1, 3, 1, 2, 1, 2, 1)\}$ in $\mathcal{H}(N)$ with zero levels, 1 and six random permutations $\{(3, 1, 1, 1, 1, 2, 2), (1, 1, 1, 1, 3, 2, 2), (1, 1, 1, 1, 2, 2, 3), (3, 2, 2, 1, 1, 1, 1), (2, 2, 3, 1, 1, 1, 1), (2, 2, 1, 1, 1, 1, 3)\}$ in $\mathcal{H}(N)$ with four levels.

4.3 Discussion

First, note that the one-to-one transformation on the index means that the above theorem can be applied to a categorical $[s]$-specified random permutation. For example, the distribution of the number of levels (homopolymers) of the (s_A, s_C, s_G, s_T)-specified DNA sequence is the same as that of the (s_1, s_2, s_3, s_4)-specified random permutation, as long as $s_A = s_1, s_C = s_2, s_G = s_3$, and $s_T = s_4$. For example, the distribution of the total number of homopolymers of length two (levels) after scanning n-nucleotides is of interest in DNA sequence analyses.

Note that for given $t, t = 1, \ldots, N, L_t(i)$ may take at most $0, 1, \ldots, s_i - 1$, and L_t may take at most $0, 1, \ldots, s_i + \cdots + s_i - i$; hence, it follows from the definitions of $s_1 N$ and $(L_t, L_t(i))$ that the size of the state space Ω_i is bounded by $\max_{1 \le t \le N} (\Omega_t) \le N s_1$. The complexity of computing the distribution of the number of levels L_N is linear rather than exponential. For moderately large n and N, the CPU time required to compute the distribution $P(L_N = \ell)$ is negligible. However, for large n and $N > 50$, the time needed to create the sequence of transition probability matrices $\{M_t\}_{t=0}^{N-1}$ can be significant.

These results show that the finite Markov chain imbedding technique and insertion procedure provide a simple, direct, and intuitive approach to studying the distributions of runs and patterns in an $[s]$-specified random permutation.

Acknowledgements This work was supported in part by the National Sciences and Engineering Research Council of Canada under Grant NSERC A-9216. The author thanks Dr. Wan-Chen Lee for her comments and suggestions. The author also thanks the editor and reviewer for their useful suggestions.

References

1. Carlitz L (1964) Extended Bernoulli and Eulerian numbers. J Duke Math 31:667–690
2. Carlitz L (1972) Enumeration of sequences by rises and falls: a refinement of the Simon Newcomb problem. J Duke Math 39:267–280
3. Carlitz L (1974) Permutations and sequences. Adv Math 14:92–120
4. Carlitz L, Scoville RA (1974) Generalized Eulerian numbers: combinatorial applications. J Reine Angew Math 265:110–137
5. David FN, Barton DE (1962) Combinatorial chance. Hafner, New York
6. Dillon JF, Roselle D (1969) Simon Newcomb's problem. SIAM J Appl Math 17:1086–1093
7. Euler L (1755) Institutiones Calculi differentialis. impensis Academiae imperialis scientiarum Petropolitanae
8. Fu JC (1995) Exact and limiting distributions of the number of successions in a random permutation. Ann Inst Statist Math 47:435–446
9. Fu JC, Koutras MV (1994) Distribution theory of runs: a Markov chain approach. J Amer Statist Assoc 89:1050–1058
10. Fu JC, Lou WYW (2000) Joint distribution of rises and falls. Ann Inst Statist Math 52:415–425
11. Fu JC, Wang LQ, Lou WYW (1999) On the exact distributions of Eulerian and Simon Newcomb numbers associated with random permutations. Statist Probab Lett 42:115–125

12. Fu JC, Lou WYW (2003) Distribution theory of runs and patterns and its applications, 1st edn. World Scientific, Singapore
13. Giladi E, Keller JB (1994) Eulerian number asymptotics. Proc Roy Soc Lond A 445:291–303
14. Harris B, Park CJ (1994) A generalization of the Eulerian numbers with a probabilistic application. Statist Probab Lett 20:37–47
15. Johnson BC (2002) The distribution of increasing 2-sequencing in random permutations of arbitrary multi-state. Statist Probab Lett 59:67–74
16. MacMahon PA (1915) Combinatory analysis. Cambridge, London
17. Nicolas JL (1992) An integral representation for Eulerian numbers. Colloq Math Soc 60:513–527
18. Riordan J (1958) An introduction to combinatorial analysis. Wiley, New York
19. Takacs L (1979) A generalization of the Eulerian numbers. Publ Math Debrecen 26:173–181
20. Tanny S (1973) A probabilistic interpretation of Eulerian numbers. J Duke Math 40:717–722

Chapter 5
Properties of General Systems of Orthogonal Polynomials with a Symmetric Matrix Argument

Yasuko Chikuse

Abstract There exists a large literature of the orthogonal polynomials (OPs) with a single variable associated with a univariate distribution. The theory of these OPs is well established and many properties of them are developed. Then, some authors have discussed the OPs with matrix arguments in the past. However, there are many unsolved properties, owing to the complex structures of the OPs with a matrix argument. In this paper, we extend some properties, which are well known for the OPs with a single variable, to those with a matrix argument. We give a brief discussion on the zonal polynomials and the general system of OPs with a symmetric matrix argument, with examples, the Hermite, the Laguerre, and the Jacobi polynomials. We derive the so-called three-term recurrence relations, and then, the Christoffel–Darboux formulas satisfied by the OPs with a symmetric matrix argument as a consequence of the three-term recurrence relations. Also, we present the "$(2k + 1)$-term recurrence relations", an extension of the three-term recurrence relations, and then an extension of the Christoffel–Darboux formulas as its consequence. Finally we give a brief discussion on the linearization problem and the representation of Hermite polynomials as moments. For the derivations of those results, the theory of zonal and invariant polynomials with matrix arguments is useful.

Keywords Christoffel–Darboux formulas · Orthogonal polynomials with a symmetric matrix argument · Symmetric matrix-variate distributions · Three-term recurrence relations · Zonal polynomials

5.1 Introduction

This paper is concerned with developing properties of the orthogonal polynomials (OPs) with a symmetric matrix argument which are associated with multivariate distributions of a symmetric matrix-variate.

Y. Chikuse (✉)
Kagawa University, Takamatsu, Japan
e-mail: ychiku0511@gmail.com

© The Author(s), under exclusive license to Springer Nature Singapore Pte Ltd. 2020   87
N. Hoshino et al. (eds.), *Pioneering Works on Distribution Theory*,
JSS Research Series in Statistics,
https://doi.org/10.1007/978-981-15-9663-6_5

There exists a large literature of the OPs with a single variable associated with a univariate distribution. The theory of these OPs is well established and many properties of them are developed. See, e.g., [1, 11, 17]. Then, some authors have discussed the OPs with matrix arguments in the past. See, e.g., [6, 7, 10, 12, 14–16]. However, there are many unsolved properties, owing to the complex structures of the OPs with matrix arguments. In this paper, we extend some properties, which are well known for the OPs with a single variable, to those with a matrix argument.

Section 5.2 discusses zonal polynomials and gives a brief discussion on the zonal polynomials and the general system of OPs with an $m \times m$ symmetric matrix argument. Examples are given, that is, the Hermite polynomials, Laguerre polynomials, and Jacobi polynomials associated with the symmetric matrix-variate normal distribution, Wishart distribution, and the matrix-variate Beta distribution, respectively. For the matrix-variate case, we discuss the so-called three-term recurrence relations in Sect. 5.3. Then, in Sect. 5.4, the Christoffel–Darboux formulas satisfied by the OPs with a symmetric matrix argument are derived, as a consequence of the above three-term recurrence relations. Here also in Sect. 5.3 we present the "$(2k + 1)-$ term recurrence relations", as an extension of the three-term recurrence relations for $k = 1$, and then an extension of the Christoffel–Darboux formulas as its consequence in Sect. 5.4. Section 5.5 gives a brief discussion on two further properties for our case of the OPs with a symmetric matrix argument, that is, the linearization problem and the representation of the Hermite polynomials with a symmetric matrix argument as moments. For the derivation of those results, the theory of zonal and invariant polynomials with matrix arguments is useful. Surely the results with $m = 1$ obtained are reduced to those for the OPs with a single variable which are already established in the literature.

5.2 Zonal Polynomials and General Systems of Orthogonal Polynomials with Matrix Arguments and Examples

5.2.1 Zonal Polynomials

We give a brief discussion on the zonal polynomials. We consider the representation of the general linear group $GL(m, R)$ of $m \times m$ real nonsingular matrices, and let $Q_l(X)$ denote the vector space of homogeneous polynomials of degree l in the elements of an $m \times m$ symmetric matrix X. The congruence transformation

$$X \longrightarrow LXL', \quad L \in GL(m, R), \tag{5.1}$$

induces the representation [2l] with the linear transformation in $Q_l(X)$. From the representation theory of $GL(m, R)$, $Q_l(X)$ decomposes into the direct sum of uniquely defined irreducible invariant subspaces

$$Q_l(X) = \bigoplus_{\lambda \vdash l} V_\lambda(X), \tag{5.2}$$

where λ runs over all ordered partitions of l into not more than m parts, denoted by $\lambda \vdash l$ for $\lambda = (l_1, \ldots, l_m), l_1 \geq \cdots \geq l_m \geq 0, \sum_{i=1}^{m} l_i = l$. When the representation is restricted to $L \in O(m)$, the orthogonal group of $m \times m$ orthonormal matrices, $V_\lambda(X)$ has a one-dimensional invariant subspace, generated by the zonal polynomials $C_\lambda(X)$.

The polynomials $C_\lambda(X)$ constitute a basis for the space of all symmetric homogeneous polynomials in the latent roots of X. See [9] ([6], Appendix A) for a more detailed discussion of zonal polynomials and invariant polynomials with symmetric matrix arguments.

5.2.2 General Systems of Orthogonal Polynomials

We consider the system of OPs $P_\lambda(X)$ with one $m \times m$ symmetric matrix argument X. Here, $P_\lambda(X)$ is invariant under the orthogonal transformations $X \longrightarrow HXH'$ for $H \in O(m)$, such that $\{P_\lambda(X)\}$ is expressed in terms of zonal polynomials and is indexed by an ordered partition.

The system of OPs has the orthogonality property

$$\int P_\lambda(X) P_\sigma(X) W(X) / [C_\lambda(I_m) C_\sigma(I_m)](dX) = h_\lambda \delta_{\lambda,\sigma}, \text{ for } \lambda \vdash l \text{ and } \sigma \vdash s$$
$$\tag{5.3}$$

with a weight function $W(X)$; we have $h_\lambda = 1$ when the OPs are normalized. Here, δ denotes the Kronecker's delta.

5.2.3 Examples

The Hermite polynomials $H_\lambda(X)$ with an $m \times m$ symmetric matrix argument X are the OPs associated with the $m \times m$ symmetric matrix-variate standard normal distribution $N_{mm}(0; I_m)$, whose density function is

$$\varphi(X) = [2^{m(m-1)/4}/(2\pi)^{m(m+1)/4}] \text{ etr } \left(-\frac{1}{2}X^2\right), \tag{5.4}$$

where etr $(A) = \exp[(tr A)]$. The orthogonality property (5.3) is specified as

$$h_\lambda = l!/C_\lambda(I_m), \tag{5.5}$$

for $P_\lambda(X) = H_\lambda(X)$ and $W(X) = \varphi(X)$.

The Laguerre polynomials $L_\lambda^u(X)$ are the OPs associated with the Wishart distribution $W_m(2u + 2p, \frac{1}{2}I_m)$, whose density function is

$$w_m(X; 2u + 2p, \frac{1}{2}I_m) = \text{etr}\,(-X) \mid X \mid^u / \Gamma_m(u + p), \quad \text{for } X > 0, \qquad (5.6)$$

where $p = (m + 1)/2$, and $\Gamma_m(a)$ is the multivariate gamma function defined by

$$\Gamma_m(a) = \int_{S>0} \text{etr}\,(-S) \mid S \mid^{a-p} (dS)$$

$$= \pi^{m(m-1)/4} \prod_{i=1}^{m} \Gamma[a - \frac{1}{2}(i - 1)]. \qquad (5.7)$$

The orthogonality property (5.3) is specified as

$$h_\lambda = l!(u + p)_\lambda / C_\lambda(I_m), \qquad (5.8)$$

for $P_\lambda(X) = L_\lambda^u(X)$ and $W(X) = w_m(X; 2u + 2p, \frac{1}{2}I_m)$, where $(a)_\lambda = \prod_{i=1}^{m}[a - \frac{1}{2}(i - 1)]_{l_i}$, for $\lambda = (l_1, \ldots l_m)$, with $(a)_l = a(a + 1) \cdots (a + l - 1)$, $(a)_0 = 1$.

See [7, 12] for the fundamental discussion of the Laguerre polynomials with a matrix argument. Chikuse ([6], Appendix B) provides a detailed discussion on $H_\lambda(X)$ and $L_\lambda^u(X)$ polynomials, i.e., the generating functions, Rodrigues formulas (differential forms), Fourier transforms, and series expressions in terms of zonal polynomials: See also [3].

The Jacobi polynomials $P_\lambda^{(a,c)}(X)$ are the OPs associated with the matrix-variate Beta distribution $B(a, c)$, whose density function is

$$\beta(X; a, c) = \{\Gamma_m(c)/[\Gamma_m(a)\Gamma_m(c - a)]\} \mid X \mid^{a-p} \mid I_m - X \mid^{c-a-p}, \quad \text{for } 0 < X < I_m. \quad (5.9)$$

A detailed discussion of Jacobi polynomials (i.e., generating function, series expressions) is provided by Davis [10]. In particular, h_λ of the orthogonality property (5.3) is provided in Davis ([10], Eq. (3.2)) for $P_\lambda(X) = P_\lambda^{(a,c)}(X)$ and $W(X) = \beta(X; a, c)$. We have the limiting relations for the three OPs

$$\begin{cases} H_\lambda(X) & = \lim_{u \to \infty} (-u^{-1/2})^l L_\lambda^u(u^{1/2}X + uI_m), \\ L_\lambda^{a-p}(X) & = \lim_{c \to \infty} P_\lambda^{(a,c)}(\frac{1}{c}X). \end{cases} \qquad (5.10)$$

5.2.4 The Extended Orthogonal Polynomials with Multiple Matrix Arguments

We can extend the OPs $\{P_\lambda(X)\}$ with one symmetric matrix argument X to $\{P_{\lambda[r];\phi}(X_{[r]})\}$ with $r(r \geq 2)$ $m \times m$ symmetric matrix arguments $X_1, \ldots, X_r (= X_{[r]})$. Here $P_{\lambda[r];\phi}(X_{[r]})$ is indexed by $\phi \in \lambda[r](\lambda_1 \cdots \cdots \lambda_r = \lambda_{[r]})$ for ordered partitions $\lambda_i \vdash l_i$ of l_i, $i = 1, \ldots, r$, and $\phi \vdash \sum_{i=1}^r l_i$, where $\phi \in \lambda[r]$ denotes that the irreducible representation $[2\phi]$ occurs in the decomposition of the Kronecker product $[2\lambda_1] \otimes \cdots \otimes [2\lambda_r]$ of the irreducible representations $[2\lambda_i]$, $i = 1, \ldots, r$. A weight function $W(X_{[r]})$ exists, such that

$$\int P_{\lambda[r];\phi}(X_{[r]}) P_{\sigma[r];\nu}(X_{[r]}) W(X_{[r]}) \prod_{i=1}^r (dX_i) = h_{\lambda[r];\phi} \delta_{\lambda[r];\phi:\sigma[r];\nu}; \qquad (5.11)$$

we have $h_{\lambda[r];\phi} = 1$ when the OPs are normalized.

The theory of invariant polynomials $C_\phi^{\lambda[r]}(X_{[r]})$ with multiple matrix arguments $X_{[r]}$, extending the zonal polynomials $C_\lambda(X)$ with one matrix argument X, is useful throughout the discussion of the polynomials with multiple matrix arguments. See, e.g., [6] and the references given therein for the discussion of the invariant polynomials: See also [2, 8, 9]. Chikuse ([6], Appendix B) gives a detailed discussion of the extended Hermite and Laguerre polynomials with multiple matrix arguments, giving those with one matrix argument as special cases: See also [4, 5]. The arguments concerning the properties of the OPs developed in this paper can be all extended to the OPs with multiple matrix arguments. However, the results obtained are complicated and lengthy, so that they are not presented in this paper.

5.3 Three-Term Recurrence Relations for $P_\lambda(X)$

5.3.1 One-Dimensional Case

We give a brief discussion of the three-term recurrence relations, which are well established on the usual one-dimensional real space.

We consider the system of OPs $\{P_n(x)\}$, with a weight function $W(x)$ such that

$$\int P_m(x) P_n(x) W(x) dx = h_n \delta_{m,n};$$

we have $h_n = 1$ when the $P_n(x)$ are normalized. Then, the system $\{P_n(x)\}$ satisfies the three-term recurrence relations:

$$x P_n(x) = a_n P_{n+1}(x) + b_n P_n(x) + c_n P_{n-1}(x), \qquad (5.12)$$

where we let $P_{-1}(x) = 0$, and a_n, b_n and c_n are real constants, and

$$a_{n-1}h_n = c_n h_{n-1}, \quad h_n = h_0 \frac{c_1 c_2 \cdots c_n}{a_0 a_1 \cdots a_{n-1}},$$

such that $a_{n-1}c_n > 0$.

5.3.2 Extension to $\{P_\lambda(X)\}$

We extend the argument to the general OPs $\{P_\lambda(X)\}$ with an $m \times m$ symmetric matrix argument X.

Theorem 5.1 *The general system $\{P_\lambda(X)\}$ satisfies the following recurrence relations:*

$$(trX) P_\lambda(X) = \sum_{\substack{\kappa \vdash (l+1)}} A_\kappa^{(\lambda)} P_\kappa(X) + \sum_{\sigma \vdash l} B_\sigma^{(\lambda)} P_\sigma(X) + \sum_{\sigma \vdash (l-1)} C_\sigma^{(\lambda)} P_\sigma(X), \quad (5.13)$$

where the coefficients A's, B's, and C's follow the conditions

$$A_\lambda^{(\tau)} h_\lambda = C_\tau^{(\lambda)} h_\tau, \text{ for } \lambda \vdash l, \ \tau \vdash (l-1), \quad (5.14)$$

and

$$B_\tau^{(\lambda)} h_\tau = B_\lambda^{(\tau)} h_\lambda, \text{ for } \lambda, \tau \vdash l. \quad (5.15)$$

Proof For a general system $\{P_\lambda(X)\}$ of OPs, let us first determine the coefficients $A_\kappa^{(\lambda)} (\kappa \vdash (l+1))$ such that $(trX)P_\lambda(X) - \sum_{\kappa \vdash (l+1)} A_\kappa^{(\lambda)} P_\kappa(X)$ is a polynomial of degree l. Then, we can write

$$(trX) P_\lambda(X) - \sum_{\kappa \vdash (l+1)} A_\kappa^{(\lambda)} P_\kappa(X) = \sum_{s=0}^{l} \sum_{\sigma \vdash s} B_\sigma^{(\lambda)} P_\sigma(X), \quad (5.16)$$

which is a polynomial of degree l, for some constants $B_\sigma^{(\lambda)}$. Multiply (5.16) by $P_\sigma(X)W(X)$ and integrate over X. Then, we have that

$$B_\sigma^{(\lambda)} = 0, \text{ for } \sigma \vdash s, s \le (l-2).$$

Thus, we can write

$$(trX)P_\lambda(X) = \sum_{\kappa \vdash (l+1)} A_\kappa^{(\lambda)} P_\kappa(X) + \sum_{\sigma \vdash l} B_\sigma^{(\lambda)} P_\sigma(X) + \sum_{\sigma \vdash (l-1)} C_\sigma^{(\lambda)} P_\sigma(X). \quad (5.17)$$

Now, multiplying (5.17) by $P_\tau(X)W(X)$ ($\tau \vdash (l-1)$) and integrating over X yields

$$\int (\text{tr}X) P_\lambda(X) P_\tau(X) W(X)(dX) = \sum_{\sigma \vdash (l-1)} C_\sigma^{(\lambda)} \int P_\sigma(X) P_\tau(X) W(X)(dX)$$

$$= C_\tau^{(\lambda)} h_\tau, \text{ where } \int P_\tau^2(X) W(X)(dX) = h_\tau. \tag{5.18}$$

Here, we have

$$(\text{tr}X) \underset{(\tau \vdash (l-1))}{P_\tau(X)} = \sum_{\kappa \vdash l} A_\kappa^{(\tau)} P_\kappa(X) + \sum_{\sigma \vdash (l-1)} B_\sigma^{(\tau)} P_\sigma(X) + \sum_{\sigma \vdash (l-2)} C_\sigma^{(\tau)} P_\sigma(X),$$

and hence

$$\int (\text{tr}X) P_\lambda(X) P_\tau(X) W(X)(dX) = \int P_\lambda(X) \sum_{\kappa \vdash l} A_\kappa^{(\tau)} P_\kappa(X) W(X)(dX)$$

$$= A_\lambda^{(\tau)} h_\lambda. \tag{5.19}$$

Combining (5.18) and (5.19) yields (5.14).

Next, multiplying (5.17) by $P_\tau(X)W(X)(\tau \vdash l)$ and integrating over X yields

$$\int (\text{tr}X) P_\lambda(X) P_\tau(X) W(X)(dX) = B_\tau^{(\lambda)} h_\tau. \tag{5.20}$$

Now, since we have

$$(\text{tr}X) \underset{(\tau \vdash l)}{P_\tau(X)} = \sum_{\kappa \vdash (l+1)} A_\kappa^{(\tau)} P_\kappa(X) + \sum_{\sigma \vdash l} B_\sigma^{(\tau)} P_\sigma(X) + \sum_{\sigma \vdash (l-1)} C_\sigma^{(\tau)} P_\sigma(X),$$

the l.h.s. of (5.20) becomes

$$B_\lambda^{(\tau)} h_\lambda. \tag{5.21}$$

Combining (5.20) and (5.21) yields (5.15).   □

Corollary 5.2 *When the $P_\lambda(X)$ are normalized such that $h_\lambda = 1$, for all $\lambda \vdash l = 0, 1, 2, \ldots$, we have the recurrence relations:*

$$(trX) \underset{(\lambda \vdash l)}{P_\lambda(X)} = \sum_{\kappa \vdash (l+1)} A_\kappa^{(\lambda)} P_\kappa(X) + \sum_{\sigma \vdash l} B_\sigma^{(\lambda)} P_\sigma(X) + \sum_{\sigma \vdash (l-1)} A_\lambda^{(\sigma)} P_\sigma(X), \tag{5.22}$$

where $B_\tau^{(\lambda)} = B_\lambda^{(\tau)}$, for $\lambda, \tau \vdash l$.

5.3.3 (2k + 1)-Term Recurrence Relations

We extend the three-term recurrence relations (5.13) with $\text{tr}X (= C_{(1)}(X))$ to "$(2k + 1)$-term recurrence relations" with $C_\kappa(X)$ in general for $\kappa \vdash k = 1, 2, 3, \ldots$.

Theorem 5.3 *Suppose that* $k \le l$. *The system* $\{P_\lambda(X)\}$ *satisfies the following recurrence relations:*

$$C_\kappa(X)P_\lambda(X) = \sum_{\sigma \vdash (l+k),\ldots,l,\ldots,(l-k)} A_\sigma^{(\lambda;\kappa)} P_\sigma(X), \tag{5.23}$$

where the coefficients A's follow the conditions:

$$A_\tau^{(\lambda;\kappa)} h_\tau = A_\lambda^{(\tau;\kappa)} h_\lambda, \text{ for } \lambda \vdash l, \ \tau \vdash (l - q), q = k, k - 1, \ldots, 0. \tag{5.24}$$

Proof With the same argument as that for Theorem 5.1, we can prove Theorem 5.3.

Now, we consider $\tau \vdash (l - q)$ for $q = k, k - 1, \ldots, 0$. Multiply (5.23) by $P_\tau(X)W(X)$ and integrate over X. Then, we have

$$\int C_\kappa(X)P_\lambda(X)P_\tau(X)W(X)(dX)$$

$$= \sum_{\sigma \vdash (l-q)} A_\sigma^{(\lambda;\kappa)} \int P_\sigma(X)P_\tau(X)W(X)(dX) = A_\tau^{(\lambda;\kappa)} h_\tau. \tag{5.25}$$

On the other hand, we have, from (5.23),

$$\underset{(\tau \vdash (l-q))}{C_\kappa(X)P_\tau(X)} = \sum_{\sigma \vdash (l-q+k),(l-q+k-1),\ldots} A_\sigma^{(\tau;\kappa)} P_\sigma(X),$$

and hence

$$\int C_\kappa(X)P_\lambda(X)P_\tau(X)W(X)(dX) = A_\lambda^{(\tau;\kappa)} h_\lambda, \tag{5.26}$$

noting that $l - q + k \ge l \ge l - q - k$.

From (5.25) and (5.26), we obtain the desired result (5.24). $\qquad\square$

Note that the case $k = 1$ reduces to Theorem 5.1.

Corollary 5.4 *When the* $P_\lambda(X)$ *are normalized such that* $h_\lambda = 1$, *we have the following recurrence relations, for* $k \le l$:

$$C_\kappa(X)P_\lambda(X) = \sum_{\sigma \vdash (l+k),(l+k-1),\ldots,l} A_\sigma^{(\lambda;\kappa)} P_\sigma(X) + \sum_{\sigma \vdash (l-1),\ldots,(l-k)} A_\lambda^{(\sigma;\kappa)} P_\sigma(X), \tag{5.27}$$

where

$$A_\tau^{(\lambda;\kappa)} = A_\lambda^{(\tau;\kappa)}, \text{ for } \lambda, \tau \vdash l. \tag{5.28}$$

5.4 Christoffel–Darboux Formula for $P_\lambda(X)$

5.4.1 Christoffel–Darboux Formulas

We consider extending the already-known result for the one-dimensional case that an important consequence of the three-term recurrence relations (5.12) is the so-called Christoffel–Darboux formulas.

Theorem 5.5 *When the $P_\lambda(X)$ are normalized such that*

$$\int P_\lambda(X) P_\kappa(X) W(X)(dX) = \delta_{\lambda,\kappa},$$

we obtain the Christoffel–Darboux formulas for $P_\lambda(X)$:

$$\sum_{l=0}^{n} \sum_{\lambda \vdash l} P_\lambda(X) P_\lambda(Y)[tr(X - Y)]$$
$$= \sum_{\lambda \vdash n} \sum_{\sigma \vdash (n+1)} A_\sigma^{(\lambda)}[P_\sigma(X) P_\lambda(Y) - P_\sigma(Y) P_\lambda(X)], \qquad (5.29)$$

where the coefficients A's are given in the three-term recurrence relations (5.13) obtained in Theorem 5.1.

Proof From the recurrence relations (5.13) given in Theorem 5.1, we have

$$(trX) P_\lambda(X) P_\lambda(Y) = \sum_{\sigma \vdash (l+1)} A_\sigma^{(\lambda)} P_\sigma(X) P_\lambda(Y)$$
$$+ \sum_{\sigma \vdash l} B_\sigma^{(\lambda)} P_\sigma(X) P_\lambda(Y) + \sum_{\sigma \vdash (l-1)} C_\sigma^{(\lambda)} P_\sigma(X) P_\lambda(Y), \qquad (5.30)$$

and

$$(trY) P_\lambda(Y) P_\lambda(X) = \sum_{\sigma \vdash (l+1)} A_\sigma^{(\lambda)} P_\sigma(Y) P_\lambda(X)$$
$$+ \sum_{\sigma \vdash l} B_\sigma^{(\lambda)} P_\sigma(Y) P_\lambda(X) + \sum_{\sigma \vdash (l-1)} C_\sigma^{(\lambda)} P_\sigma(Y) P_\lambda(X). \qquad (5.31)$$

Subtracting (5.31) from (5.30) and then summing over $\lambda \vdash l$ yields

$$\sum_{\lambda \vdash l} P_\lambda(X) P_\lambda(Y)[\mathrm{tr}(X - Y)]$$

$$= \sum_{\lambda \vdash l} \sum_{\sigma \vdash (l+1)} A_\sigma^{(\lambda)}[P_\sigma(X) P_\lambda(Y) - P_\sigma(Y) P_\lambda(X)]$$

$$+ \sum_{\lambda \vdash l} \sum_{\sigma \vdash l} B_\sigma^{(\lambda)}[P_\sigma(X) P_\lambda(Y) - P_\sigma(Y) P_\lambda(X)]$$

$$+ \sum_{\lambda \vdash l} \sum_{\sigma \vdash (l-1)} C_\sigma^{(\lambda)}[P_\sigma(X) P_\lambda(Y) - P_\sigma(Y) P_\lambda(X)]. \tag{5.32}$$

Now, under the assumption of the normalization of $P_\lambda(X)$, we have

$$C_\sigma^{(\lambda)} = A_\lambda^{(\sigma)}, \text{ for } \lambda \vdash l, \sigma \vdash (l-1),$$

and

$$B_\sigma^{(\lambda)} = B_\lambda^{(\sigma)}, \text{ for } \lambda, \sigma \vdash l.$$

Then, the second term of (5.32) becomes

$$\sum_{\lambda, \sigma \vdash l} B_\sigma^{(\lambda)} P_\sigma(X) P_\lambda(Y) - \sum_{\lambda, \sigma \vdash l} B_\lambda^{(\sigma)} P_\lambda(Y) P_\sigma(X)$$

$$= \sum_{\lambda, \sigma \vdash l} (B_\sigma^{(\lambda)} - B_\lambda^{(\sigma)}) P_\sigma(X) P_\lambda(Y) = 0.$$

Then, summing (5.32) over $l = 1, \ldots, n$, we have the l.h.s of the resulting equation

$$\sum_{l=1}^{n} \sum_{\lambda \vdash l} P_\lambda(X) P_\lambda(Y)[\mathrm{tr}(X - Y)], \tag{5.33}$$

and the r.h.s of the resulting equation

$$\sum_{l=1}^{n} \sum_{\lambda \vdash l} \sum_{\sigma \vdash (l+1)} A_\sigma^{(\lambda)}[P_\sigma(X) P_\lambda(Y) - P_\sigma(Y) P_\lambda(X)]$$

$$- \sum_{l=1}^{n} \sum_{\lambda \vdash (l-1)} \sum_{\sigma \vdash l} A_\sigma^{(\lambda)}[P_\lambda(Y) P_\sigma(X) - P_\lambda(X) P_\sigma(Y)]$$

$$= \sum_{\lambda \vdash n} \sum_{\sigma \vdash (n+1)} A_\sigma^{(\lambda)}[P_\sigma(X) P_\lambda(Y) - P_\sigma(Y) P_\lambda(X)]$$

$$- A_{(1)}^{(0)}[P_{(1)}(X) P_{(0)}(Y) - P_{(1)}(Y) P_{(0)}(X)]. \tag{5.34}$$

Note that, under the assumption $h_\lambda = 1$, $P_{(0)}(X) = \pm 1$, $P_{(-1)}(X) = 0$, and, from (5.29) with $l = 0$,

$$(\operatorname{tr}X)P_{(0)}(X) = A^{(0)}_{(1)}P_{(1)}(X) + B^{(0)}_{(0)}P_{(0)}(X),$$

and hence

$$\operatorname{tr}(X - Y)P_{(0)}(X) = A^{(0)}_{(1)}(P_{(1)}(X) - P_{(1)}(Y)),$$

that is,

$$P_{(0)}(X)P_{(0)}(Y)\operatorname{tr}(X - Y) = A^{(0)}_{(1)}[P_{(1)}(X)P_{(0)}(Y) - P_{(1)}(Y)P_{(0)}(X)]. \qquad (5.35)$$

Combining (5.33)–(5.35), we obtain the desired result (5.29). $\qquad\square$

Note Putting $m = 1$ in Theorem 5.5, we obtain the following known result. Suppose that the OPs $P_n(x)$ with a single variable x are normalized such that $h_n = 1$, $n = 0, 1, \ldots$. Then, we have

$$\sum_{l=0}^{n} P_l(x)P_l(y) = a_n \frac{P_{n+1}(x)P_n(y) - P_{n+1}(y)P_n(x)}{x - y}, \qquad (5.36)$$

where a_n is given in the three-term recurrence relations (5.12).

5.4.2 Extending the Christoffel–Darboux Formulas

We now extend the Christoffel–Darboux formulas (5.29) so that we replace $\operatorname{tr}X = C_{(1)}(X)$ with more general $C_\kappa(X)$, for $\kappa \vdash k = 1, 2, 3, \ldots$.

Theorem 5.6 *Suppose that $k \le l$ and that the $P_\lambda(X)$ are normalized such that $h_\lambda = 1$ for all λ. Then, we obtain the following formulas for $\{P_\lambda(X)\}$, for $\kappa \vdash k$:*

$$\sum_{l=0}^{n} \sum_{\lambda \vdash l} P_\lambda(X)P_\lambda(Y)[C_\kappa(X) - C_\kappa(Y)]$$

$$= \sum_{q=0}^{k-1} \sum_{\substack{\lambda \vdash (n-q) \\ \sigma \vdash (n-q+k),\ldots,(n+1)}} A^{(\lambda;\kappa)}_\sigma [P_\sigma(X)P_\lambda(Y) - P_\sigma(Y)P_\lambda(X)]. \quad (5.37)$$

Here, we define $P_\tau(X) = 0$, for $\tau \vdash -1, -2, -3, \ldots$, and the coefficients A's are given in the "$(2k + 1)$-term recurrence relations" (5.23) obtained in Theorem 5.3.

Proof From the $(2k + 1)$-term recurrence relations (5.23) given in Theorem 5.3, we have

$$C_\kappa(X)P_\lambda(X)P_\lambda(Y) = \sum_{\sigma \vdash (l+k),\ldots,(l-k)} A^{(\lambda;\kappa)}_\sigma P_\sigma(X)P_\lambda(Y), \qquad (5.38)$$

and

$$C_\kappa(Y)P_\lambda(Y)P_\lambda(X) = \sum_{\sigma \vdash (l+k),\ldots,(l-k)} A_\sigma^{(\lambda;\kappa)} P_\sigma(Y)P_\lambda(X), \tag{5.39}$$

$$\text{where } A_\sigma^{(\lambda;\kappa)} = A_\lambda^{(\sigma;\kappa)}, \quad \text{for } \lambda \vdash l, \ \sigma \vdash l, (l-1), \ldots, (l-k).$$

Subtracting (5.39) from (5.38) and then summing over $\lambda \vdash l$ yields

$$\sum_{\lambda \vdash l} P_\lambda(X)P_\lambda(Y)[C_\kappa(X) - C_\kappa(Y)]$$

$$= \sum_{\substack{\lambda \vdash l \\ \sigma \vdash (l+k),\ldots,l,\ldots(l-k)}} A_\sigma^{(\lambda;\kappa)}[P_\sigma(X)P_\lambda(Y) - P_\sigma(Y)P_\lambda(X)]. \tag{5.40}$$

Now, similarly to the proof of Theorem 5.5, we can show that

$$\sum_{\substack{\lambda \vdash l \\ \sigma \vdash l}} A_\sigma^{(\lambda;\kappa)}[P_\sigma(X)P_\lambda(Y) - P_\sigma(Y)P_\lambda(X)] = 0.$$

Then, summing (5.40) over $l = 1, \ldots, n$, we have the l.h.s. of the resulting equation

$$\sum_{l=1}^{n} \sum_{\lambda \vdash l} P_\lambda(X)P_\lambda(Y)[C_\kappa(X) - C_\kappa(Y)], \tag{5.41}$$

and the r.h.s. of the resulting equation

$$\sum_{l=1}^{n} \sum_{\substack{\lambda \vdash l \\ \sigma \vdash (l+k),\ldots,(l+1)}} A_\sigma^{(\lambda;\kappa)}[P_\sigma(X)P_\lambda(Y) - P_\sigma(Y)P_\lambda(X)]$$

$$+ \sum_{l=1}^{n} \sum_{\substack{\sigma \vdash l \\ \lambda \vdash (l-1),\ldots,(l-k)}} A_\lambda^{(\sigma;\kappa)}[P_\lambda(X)P_\sigma(Y) - P_\lambda(Y)P_\sigma(X)],$$

$$\text{with } A_\lambda^{(\sigma;\kappa)} = A_\sigma^{(\lambda;\kappa)},$$

$$= \left\{ \sum_{l=1}^{n} \sum_{\substack{\lambda \vdash l \\ \sigma \vdash (l+k),\ldots,(l+1)}} - \sum_{l=1}^{n} \sum_{\substack{\lambda \vdash (l-1),\ldots,(l-k) \\ \sigma \vdash l}} \right\}$$

$$\times A_\sigma^{(\lambda;\kappa)}[P_\sigma(X)P_\lambda(Y) - P_\sigma(Y)P_\lambda(X)]. \tag{5.42}$$

Now, we consider the implications of (5.42). Note that under the assumption $h_\lambda = 1$, $\lambda \vdash 0, 1, 2, \ldots$, $P_{(0)}(X) = \pm 1$, and that, from (5.23) with $l = 0$ (i.e., for $\lambda = (0)$),

$$C_\kappa(X)P_{(0)}(X) = \sum_{\sigma \vdash k,(k-1),\ldots,0} A_\sigma^{(0;\kappa)} P_\sigma(X),$$

and hence

$$[C_\kappa(X) - C_\kappa(Y)]P_{(0)}(X) = \sum_{\sigma \vdash k,(k-1),...,1} A_\sigma^{(0;\kappa)}(P_\sigma(X) - P_\sigma(Y)), \qquad (5.43)$$

where we note that

$$P_\tau(X) = 0, \ \tau \vdash -1, -2, \dots.$$

After a "careful consideration", we see that (5.42) is simplified as

$$\sum_{q=0}^{k-1} \sum_{\substack{\lambda \vdash (n-q) \\ \sigma \vdash (n-q+k),...,(n+1)}} A_\sigma^{(\lambda;\kappa)}[P_\sigma(X)P_\lambda(Y) - P_\sigma(Y)P_\lambda(X)]$$

$$- \sum_{\sigma \vdash k,(k-1),...,1} A_\sigma^{(0;\kappa)}[P_\sigma(X)P_{(0)}(Y) - P_\sigma(Y)P_{(0)}(X)]. \qquad (5.44)$$

Combining (5.41)–(5.44), we obtain the desired result (5.37).

Note that (5.37) reduces to the Christoffel–Darboux formulas given in Theorem 5.5, for $k = 1$ (i.e., for $\kappa = (1)$).

5.5 Other Properties

In this section, we briefly discuss two additional properties of our OPs $P_\lambda(X)$.

5.5.1 Linearization Problem

Suppose that we have OPs $P_\lambda(X)$ with a weight function $W(X)$ such that we have the orthogonality property (5.3)

$$\int \frac{P_\kappa(X)}{C_\kappa(I_m)} \frac{P_\lambda(X)}{C_\lambda(I_m)} W(X)(dX) = h_\kappa \delta_{\kappa,\lambda}.$$

When we write the product of the OPs as

$$\frac{P_\lambda(X)P_\nu(X)}{C_\lambda(I_m)C_\nu(I_m)} = \sum_{\sigma \vdash s} C_\sigma(\lambda, \nu) \frac{P_\sigma(X)}{C_\sigma(I_m)}, \qquad (5.45)$$

the problem becomes one of determining the coefficients $C_\sigma(\lambda, \nu)$.

Now, if we can evaluate the integral

$$\int \frac{P_\kappa(X) P_\lambda(X) P_\nu(X)}{C_\kappa(I_m) C_\lambda(I_m) C_\nu(I_m)} W(X)(dX) = d(\kappa, \lambda, \nu), \text{ say,}$$

then, we can write

$$d(\kappa, \lambda, \nu) = \int \frac{P_\kappa(X)}{C_\kappa(I_m)} [\sum_{\sigma \vdash s} C_\sigma(\lambda, \nu) \frac{P_\sigma(X)}{C_\sigma(I_m)}] W(X)(dX)$$

$$= \sum_{\sigma \vdash s} C_\sigma(\lambda, \nu) h_\kappa \delta_{\kappa, \sigma} = C_\kappa(\lambda, \nu) h_\kappa.$$

Thus, we obtain

$$C_\kappa(\lambda, \nu) = d(\kappa, \lambda, \nu) / h_\kappa.$$

Example

For the Hermite polynomials $P_\lambda(X) = H_\lambda(X)$ with $W(X) = \varphi(X)$, the orthogonality property gives $h_\kappa = k! / C_\kappa(I_m)$, as in (5.3), while some further algebraic calculation yields

$$d(\kappa, \lambda, \nu) = \frac{k! \, l! \, n! \, \Psi_{\kappa, \lambda, \nu}}{\left(\frac{k+l-n}{2}\right)! \left(\frac{l+n-k}{2}\right)! \left(\frac{k+n-l}{2}\right)!}, \tag{5.46}$$

where we follow the conditions

$$k + l + n, \text{ even integers and}$$
$$k + l \geq n, \ l + n \geq k \text{ and } k + n \geq l. \tag{5.47}$$

Here, the expression of the coefficient $\Psi_{\kappa, \lambda, \nu}$ can be evaluated using the theory of invariant polynomials but complicated and lengthy, and hence is omitted here. Note that the case $m = 1$ yields $\Psi = 1$, and thus we obtain the known result for the linearization of the Hermite polynomials with a single variable.

5.5.2   *Representation of the $H_\lambda(X)$ as Moments*

Proposition We can express the Hermite polynomials $H_\lambda(X)$ as moments; that is,

$$H_\lambda(X) = E_Y[C_\lambda(X + iY)], \tag{5.48}$$

where Y is distributed as symmetric matrix-variate standard normal $N_{mm}(0; I_m)$.

Proof We write the generating function of $E_Y[C_\lambda(X + iY)]$ as

$$g \cdot f = \sum_{\substack{0 \\ \lambda \vdash l}}^{\infty} E_Y[C_\lambda(X + iY)]C_\lambda(T)/[C_\lambda(I_m)l!]$$

$$= E_Y \int_{H \in O(m)} \mathrm{etr}[(X + iY)H'TH][dH]$$

$$= \int_{H \in O(m)} \mathrm{etr}(XH'TH)[E_Y \mathrm{etr}(iYH'TH)][dH],$$

where $[dH]$ is the normalized invariant measure on $O(m)$, using the theory of zonal polynomials of [13]. Thus, we can write

$$g \cdot f = \int_{H \in O(m)} \mathrm{etr}(XH'TH)\mathrm{etr}(-\frac{1}{2}T^2)[dH],$$

which is the generating function for the Hermite polynomials $\{H_\lambda(X)\}$. □

Various properties of the Hermite polynomials $H_\lambda(X)$ are obtained from (5.48). Note that when $m = 1$, (5.48) reduces to the results due to [18, 19].

Acknowledgements I would like to express my sincere thanks to Professors N. Hoshino and Y. Yokoyama for assisting with the typesetting of the manuscript. I also appreciate Professor Shibuya's comments and suggestions concerning the orthogonal polynomials. The author is grateful to the referee for invaluable comments and suggestions, which led to an improved version of this paper.

References

1. Andrews GE, Askey R, Roy R (1999) Special functions, vol 71. Encyclopedia of mathematics and its applications. Cambridge University Press, Cambridge
2. Chikuse Y (1980) Invariant polynomials with matrix arguments and their applications. In: Gupta RP (ed) Multivariate statistical analysis. North-Holland, Amsterdam, pp 53–68
3. Chikuse Y (1992a) Properties of Hermite and Laguerre polynomials in matrix argument and their applications. Linear Algebra Appl 176:237–260
4. Chikuse Y (1992b) Generalized Hermite and Laguerre polynomials in multiple symmetric matrix arguments and their applications. Linear Algebra Appl 176:261–287
5. Chikuse Y (1994) Generalized noncentral Hermite and Laguerre polynomials in multiple matrices. Linear Algebra Appl 210:209–226
6. Chikuse Y (2003) Statistics on special manifolds, vol 174. Lecture notes in statistics. Springer, New York
7. Constantine AG (1966) The distribution of Hotelling's generalized T_0^2. Ann Math Statist 37:215–225
8. Davis AW (1979) Invariant polynomials with two matrix arguments extending the zonal polynomials: applications to multivariate distribution theory. Ann Inst Statist Math 31(A):465–485
9. Davis AW (1980) Invariant polynomials with two matrix arguments, extending the zonal polynomials. In: Krishnaiah PR (ed) Multivariate analysis, vol V. North-Holland, Amsterdam, pp 287–299
10. Davis AW (1999) Special functions on the Grassmann manifold and generalized Jacobi polynomials, Part I. Linear Algebra Appl 289:75–94

11. Erdélyi A, Magnus W, Oberhettinger F, Tricomi FG (1953) Higher transcendental functions, vol II. McGraw-Hill, New York
12. Herz CS (1955) Bessel functions of matrix argument. Ann Math 61:474–523
13. James AT (1964) Distributions of matrix variates and latent roots derived from normal samples. Ann Math Statist 35:475–501
14. James AT (1976) Special functions of matrix and single argument in statistics. In: Askey RA (ed) Theory and applications of special functions. Academic Press, New York, pp 497–520
15. James AT, Constantine AG (1974) Generalized Jacobi polynomials as special functions of the Grassmann manifold. Proc Lond Math Soc 28:174–192
16. Muirhead RJ (1982) Aspects of multivariate statistical theory. Wiley, New York
17. Szegö G (1975) Orthogonal polynomials, 4th edn. American Mathematical Society, Providence, R.I
18. Willink R (2005) Normal moments and Hermite polynomials. Stat Probabil Lett 73:271–275
19. Withers CS (2000) A simple expression for the multivariate Hermite polynomials. Stat Probabil Lett 47:165–169

Chapter 6
A Characterization of Jeffreys' Prior with Its Implications to Likelihood Inference

Takemi Yanagimoto and Toshio Ohnishi

Abstract A characterization of Jeffreys' prior for a parameter of a distribution in the exponential family is given by the asymptotic equivalence of the posterior mean of the canonical parameter to the maximum likelihood estimator. A promising role of the posterior mean is discussed because of its optimality property. Further, methods for improving estimators are explored, when neither the posterior mean nor the maximum likelihood estimator performs favorably. The possible advantages of conjugate analysis based on a suitably chosen prior are examined.

Keywords Asymptotic equivalence · Bayesian and frequentist compromise · Characterization of a prior · MLE · Optimal predictor · Optimality of an estimator · Reference prior

6.1 Introduction

An optimal predictor was first introduced by Aitchison [1] as the posterior mean of future densities $p(y|\theta)$, $\theta \in \Theta$. His introduction is explained in a general way by applying the differential geometric notion of the α-connection; see [2] for this notion. Corcuera and Giummole [5] generalized this to a family of predictors indexed by α. Each predictor is optimal under the corresponding α-divergence loss. There are two typical divergences referred to as e- and m-divergences. The predictor given by Aitchison [1] satisfies an optimality property associated with m-divergence. The optimal predictor associated with e-divergence was further explored by Yanagimoto and Ohnishi [21], leading to the e-optimal predictor. A useful property for our later discussions is that this can be expressed as a plug-in predictor in terms of the posterior

T. Yanagimoto (✉)
Institute of Statistical Mathematics, 10-3 Midori-cho, Tachikawa, Tokyo 190-8562, Japan
e-mail: yanagmt@ism.ac.jp

T. Ohnishi
Faculty of Economics, Kyushu University, 744 Motooka Nishi-ku, Fukuoka 819-0395, Japan
e-mail: ohnishi@econ.kyushu-u.ac.jp

103
N. Hoshino et al. (eds.), *Pioneering Works on Distribution Theory*,
JSS Research Series in Statistics,
https://doi.org/10.1007/978-981-15-9663-6_6

mean of the canonical parameter θ. This fact connects an estimator with the plug-in predictor in terms of the estimate and allows us to develop conjugate analysis based on the optimal predictor. Although we suggest the possible usefulness of the procedures induced from the optimal predictor, there remain many to be dissolved before recommending actual applications of them.

Let $x = (x_1, \ldots, x_n)$ be a sample vector of size n from a population with a density function $p(x|\theta)$, and let y be an unobserved (future) sample that is independently and identically distributed. When the sampling density is in the exponential family of the form $p(x|\theta) = \exp\{n(\bar{t}\theta - M(\theta))\}a(x)$, conjugate priors provide us with useful Bayesian models. Explicit expressions are possible, and they make our understanding of the induced procedures easier. First, we claim that a predictor plays an important role in discussing the estimation of a parameter. This is because our naive understanding of an estimate $\check{\theta}$ is obtained by the plug-in predictor of the estimate $p(y|\check{\theta})$.

Likelihood inference is still popular, and the maximum likelihood estimator (MLE) $\hat{\theta}_{ML}$ and the likelihood ratio test statistic are routinely employed in various fields of practical applications. Thus, it is worth demonstrating the advantages of a Bayesian method under various conditions. We show that the posterior mean of the canonical parameter under Jeffreys' prior density $\hat{\theta}$ is asymptotically equivalent to the MLE up to the order $O(1/n^2)$. Conversely, this asymptotic statement characterizes Jeffreys' prior density. The maximized likelihood $p(x|\hat{\theta}_{ML})$ is then expected to be close to the observed predictor $p(x|\hat{\theta})$, which allows us to explain the likelihood ratio test statistic using Bayesian terms. An important point in this study is that neither the MLE nor the posterior mean of the canonical parameter is necessarily recommended in wide ranges of practical applications. In such cases, we recommend the choice of a prior density more suitable over Jeffreys' prior density.

The present paper is constructed as follows. Section 6.2 reviews related subjects to provide a theoretical background on conjugate analysis based on Jeffreys' prior density. The main theorem presented in Sect. 6.3 characterizes Jeffreys' prior in relation to the asymptotic behavior of the posterior mean; mathematical proofs are given in the appendices. Section 6.4 presents procedures induced from the posterior mean under Jeffreys' prior. The potential advantages of Bayesian procedures are discussed from their optimality property and the flexible choice of a suitable prior in Sect. 6.5. We close this paper by summarizing the recommendations based on this study.

6.2 Preliminaries

Consider a family of sampling densities in the regular exponential family of the form

$$\mathcal{F} = \left\{ p(x|\theta) = \exp\{n(\bar{t}\theta - M(\theta))\}a(x)|\theta \in \Theta \right\} \tag{6.1}$$

with $\bar{t} = \sum t_i/n$. The parameter θ is called the canonical parameter. We usually assume that the dimension of Θ is p, but for notational simplicity, we use the scalar notation unless any confusion is anticipated. We do not pursue mathematically rigorous expressions, except for the proof of the theorem. Instead, we focus our attention on inferential procedures.

The mean parameter $\mu = E\{t; p(x|\theta)\}$) is expressed as $M'(\theta)$, where $E\{g(z); f(z)\}$ denotes the expectation of $g(z)$ under a probability density $f(z)$. Write the dual convex function to $M(\theta)$ as $N(\mu)$. Then, it holds that $\theta = N'(\mu)$. The conjugate prior density with a supporting density/function $b(\theta)$ is of the form

$$\pi(\theta; m, \delta) = \exp\{\delta(m\theta - M(\theta) - N(m))\} b(\theta)k(m, \delta). \qquad (6.2)$$

We allow δ to take the value zero, even when $b(\theta)$ is improper, if the posterior density is defined for every x. For convenience, we call this the supporting density even when it is improper. Our primary interest is in the choice of the supporting density.

The posterior density, induced from the sampling density (6.1) and the prior density (6.2), is expressed as

$$\pi(\theta|x) = \exp\{(n+\delta)(\mu^*\theta - M(\theta) - N(\mu^*))\} b(\theta)k(\mu^*, n+\delta) \qquad (6.3)$$

with $\mu^* = (n\bar{t} + \delta m)/(n + \delta)$. We employ the less familiar notation μ^* to emphasize that this statistic is not necessarily an estimator of μ. The form of the posterior density is given by replacing m and δ in a prior density (6.2) by μ^* and $n + \delta$, respectively. This property is called *closed under the sampling*, and the notion is the key in conjugate analysis.

6.2.1 Jeffreys' Prior in Conjugate Analysis

A Bayesian model is specified by choosing a value of δ and assuming a supporting density $b(\theta)$, which is often referred to as a non-informative prior density. One of familiar prior densities is due to [12], which is written as

$$\pi_J(\theta) \propto \sqrt{M''(\theta)}. \qquad (6.4)$$

The posterior density based on this prior density is written as $\pi_J(\theta|x)$. A favorable property of this prior density is its invariance under a parameter transformation. Another naive prior density is the uniform prior density, $\pi_U(\theta) \propto 1$. The conjugate prior densities (6.2) based on these prior densities are written as $\pi_J(\theta; m, \delta)$ and $\pi_U(\theta; m, \delta)$, respectively. The other familiar prior density is the reference prior density due to [3, 4].

6.2.2 Kullback–Leibler Divergence

The Kullback–Leibler divergence between two densities $p_1(y)$ and $p_2(y)$, $D(p_1(y),$ $p_2(y))$, is expressed as $E\{\log\big(p_1(y)/\ p_2(y)\big); p_1(y)\}$. Setting $p_1(y) = p(y|x)$ and $p_2(y) = p(y|\theta)$, we can define the divergence between the two predictors $p(y|x)$ and $p(y|\theta)$ as $D(p(y|x), p(y|\theta))$. Accordingly, setting $p_1(y) = p(y|\theta')$ and $p_2(y) = p(y|\theta)$, we can define the divergence between the parameters θ' and θ as $D(\theta', \theta) = D(p(y|\theta'), p(y|\theta))$. When the sampling density is in the exponential family (6.1), the divergence $D(\theta', \theta)$ is explicitly written as $n(N(\theta') + M(\theta) - \mu'\theta)$, where μ' is the mean parameter corresponding to θ', that is, $\mu' = M'(\theta')$. The dual divergence is defined by $D(\theta, \theta')$. These divergences play vital roles in conjugate analysis. The prior density (6.2) is rewritten as $\pi(\theta; m, \delta) = \exp-\{\delta D(c, \theta)\}b(\theta)k(m, \delta)$ with $c = N'(m)$.

These dual divergences define the dual losses of a predictor $p(y|x)$ as $D(p(y|x),$ $p(y|\theta))$ and $D(p(y|\theta), p(y|x))$, respectively. They are called the e-divergence loss and the m-divergence loss, respectively, because they are associated with the e- and m-geodesics in the differential geometry of statistics, respectively. Accordingly, the dual divergences define the dual losses of an estimator $\check{\theta}$ as $D(\check{\theta}, \theta) (=$ $D(p(y|\check{\theta})), p(y|\theta))$ and $D(\theta, \check{\theta})$. We will discuss the estimation of a parameter based on the optimum predictor under the e-divergence loss. The nomenclature of the e-optimal predictor comes from the e-divergence loss.

6.2.3 Optimal Predictors

The e-optimal predictor is given by

$$p_e(y|x) \propto \exp\big(E\{\log p(y|\theta); \pi(\theta|x)\}\big). \qquad (6.5)$$

This predictor was introduced as a member of the family of α-optimal predictor by Corcuera and Giummole [5], which was explored further by Yanagimoto and Ohnishi [21, 22]. This optimal predictor minimizes the posterior mean of the e-divergence loss $D(p(y|x), p(y|\theta))$. When the sampling density is in the exponential family, it is expressed as $p_e(y|x) = p(y|\hat{\theta})$ in terms of the posterior mean of the canonical parameter θ. Since it holds that $p_e(y|x) \in \mathcal{F}$, we observe another closedness property between the sampling density and the optimal predictor.

The normalizing constant of the e-optimal predictor in (6.5) is expressed in terms of the e-divergence loss, and so the e-optimal predictor is explicitly written as

$$p_e(y|x) = \exp\big(E\{\log p(y|\theta) + D(\hat{\theta}, \theta); \pi(\theta|x)\}\big). \qquad (6.6)$$

Next, we discuss two equalities that are useful for evaluating the posterior mean of θ, $\hat{\theta}$, and the MLE of θ, $\hat{\theta}_{ML}$. The MLE of μ is expressed as $\hat{\mu}_{ML} = \bar{t}$, and the relationships, $\hat{\mu}_{ML} = M'(\hat{\theta}_{ML})$ and $\hat{\theta}_{ML} = N'(\hat{\mu}_{ML})$, hold between them. The following well-known equality describes the sample-wise relationship between the logarithmic likelihood ratio and e-divergence.

$$\log\left\{\frac{p(x|\hat{\theta}_{ML})}{p(x|\theta)}\right\} = D\big(p(y|\hat{\theta}_{ML}),\ p(y|\theta)\big) \tag{6.7}$$

when the sampling density is in the exponential family. The corresponding equality follows from (6.6).

$$E\left\{\log\frac{p_e(x\,|\,x)}{p(x|\theta)};\ \pi(\theta|x)\right\} = E\{D\big(p_e(y\,|\,x),\ p(y|\theta)\big);\ \pi(\theta|x)\}.$$

These two equalities indicate the severe limitation of the use of the MLE. Indeed, the maximized likelihood on the left-hand side of (6.7) is directly associated with the excess of the e-divergence loss. In addition, a suitable method for reducing the posterior mean of the e-divergence loss is to choose a Bayesian estimator so that the observed e-optimal predictor can be as small as possible.

The dual optimal predictor in the family, introduced by Aitchison [1], is expressed as $p_m(y|x) = E\{p(y|\theta);\ \pi(\theta|x)\}$. This predictor minimizes the posterior mean of the m-divergence loss $D(p(y|\theta),\ p(y|x))$.

6.3   A Characterization of Jeffreys' Prior

A notable fact is that the posterior mean of the canonical parameter θ under Jeffreys' prior function is asymptotically equivalent to the MLE of θ, which is also the posterior mode under the uniform prior function for θ.

Before stating this assertion formally, we prepare a characterization theorem for Jeffreys' prior function. In this section, we explicitly treat the parameter θ as a p-dimensional vector, and the gradient of $g(\theta)$ is denoted by $\nabla g(\theta)$. Further in this section, we will distinguish rigorously a prior density from a prior function.

Theorem 6.1 *Suppose that the cumulant function $M(\theta)$ is C^4-class differentiable, and that the Fisher information matrix $I(\theta)$ is positive definite. Let $b(\theta)$ be a C^2-class differentiable non-negative function that defines the prior density*

$$\pi(\theta;\theta_0, n) \propto \exp\{-nD(\theta_0, \theta)\}b(\theta).$$

Then a necessary and sufficient condition for the asymptotic relationship

$$E\{\theta;\ \pi(\theta;\theta_0, n)\} = \theta_0 + O(n^{-2})$$

for every $\theta_0 \in \Theta$ is that $b(\theta)$ is the solution of the partial differential equation

$$\nabla \left\{ \log \frac{b(\theta)}{\sqrt{\det I(\theta)}} \right\} = 0. \tag{6.8}$$

The proof is given in Appendix A. Jeffreys' prior function is the unique solution of (6.8).

Note that the statement in Theorem 6.1 holds under a true θ_0, which makes rigorous treatment possible. The proof is based on the Laplace approximation. Similar techniques necessary for Bayesian estimation are discussed in [20]. A close result is seen in [9]. They expanded formally the posterior mean around the MLE and provided only the part of sufficiency in a less rigorous way. We present the uniqueness property under the relaxed regularity conditions on the necessary order of differentiability of $M(\theta)$ and $b(\theta)$. Further, the order of convergence is evaluated explicitly.

Our aim here is to examine the asymptotic relationship of the posterior mean of θ with the MLE of θ, which is written as $\nabla N(\bar{t})$. The posterior density $\pi(\theta|x)$ under Jeffreys' prior function is expressed as $\pi_J(\theta; \hat{\theta}_{ML}, n)$. Thus, if we can replace θ_0 in Theorem 6.1 by $\hat{\theta}_{ML}$, the required result follows. The following corollary validates this formal replacement:

Corollary 6.1 *Under the regularity conditions in Theorem 6.1, the posterior mean of the canonical parameter under Jeffreys' prior density is asymptotically equivalent in probability to the MLE up to the order $O(n^{-2})$, that is,*

$$E\{\theta; \pi_J(\theta|x)\} = \hat{\theta}_{ML} + O_P(n^{-2}). \tag{6.9}$$

A brief proof is given in Appendix B.

Corollary 6.1 can be extended to the general case of a conjugate prior density $\pi_J(\theta; m, \delta)$. The induced posterior density becomes $\pi_J(\theta; \mu^*, n + \delta)$. Note that $\theta^* = \nabla N(\mu^*)$ is the posterior mode of θ under a prior function $\pi_U(\theta)$.

Corollary 6.2 *The posterior mean of the canonical parameter under a conjugate prior function based on Jeffreys' prior density is asymptotically equivalent in probability up to the order $O(n^{-2})$ to $\theta^* = \nabla N(\mu^*)$, that is,*

$$E\{\theta; \pi_J(\theta|x)\} = \theta^* + O_P(n^{-2}). \tag{6.10}$$

As pointed out by Diaconis and Ylvisaker [7], a regularity condition is necessary so that the posterior mean of μ under the conjugate prior density $\pi_U(\theta; m, \delta)$ becomes μ^*. As a special case, the MLE of μ becomes $\bar{t}$, which is the posterior mean of μ under $\pi_U(\theta)$. In this sense, the estimator μ^* in traditional conjugate analysis is regarded as an extension of the MLE.

The following corollary concerns the normalizing constant $k(\mu^*, n)$ of a prior density of the form in (6.2). In some familiar conjugate priors for parameters, such as the mean parameter in the normal distribution and the mean parameter in the exponential distribution, $k(\mu^*, n)$ is independent of μ^*. Differentiating the equality $\int \pi(\theta|x)dx = 1$ for the posterior density (6.3) with respect to μ^*, we obtain the following equation

$$(n + \delta)\{E\{\theta; \pi(\theta \,|\, x)\} - \nabla N(\mu^*)\} - \frac{\partial}{\partial \mu^*} \log k(\mu^*, n + \delta) = 0. \qquad (6.11)$$

Applying this equality and Corollary 6.2, we immediately have the following.

Corollary 6.3 *Let* $\pi(\theta; m, \delta)$ *be a conjugate prior function. Then the normalizing constant* $k(\mu^*, n)$ *satisfies the asymptotic relationship.*

$$\frac{\partial}{\partial \mu^*} \log k(\mu^*, n + \delta) = O_P\left(\frac{1}{n}\right)$$

if the asymptotic equality (6.10) *holds.*

6.4   Implications to Likelihood Inferential Procedures

Various Bayesian procedures corresponding to likelihood inferential ones are formally induced from the asymptotic equivalence given in the previous section. Assume that the sampling density (6.1) is in the exponential family, and that a conjugate prior density is of the form (6.2). We first review the e-optimal predictor based on Jeffreys' prior density $b(\theta) = \pi_J(\theta)$. Since the e-optimal predictor (6.5) is expressed as a plug-in predictor, we can simultaneously define the estimator and the predictor, respectively, as

$$\hat{\theta} = E\{\theta; \pi_J(\theta | x)\} \qquad (6.12)$$

and

$$p(y | x) = p_e(y | x) = p(y | \hat{\theta}). \qquad (6.13)$$

A transformed parameter $\xi = g(\theta)$ is estimated by $\hat{\xi} = g(\hat{\theta})$, when the transformation is suitably defined. Corollary 6.2 indicates that this estimator and the MLE are close to each other, when n is large.

The asymptotic equivalence between the observed predictor $p(x | \hat{\theta})$ and the observed plug-in predictor in terms of the MLE $p(x | \hat{\theta}_{ML})$ allows us to use the observed optimal predictor to define the likelihood ratio statistic. Writing $\theta = (\theta_1, \theta_2)$, we consider the subfamily $p(x | (\theta_1, c_2))$ for a known value c_2 of θ_2 in the exponential family. Then, the log-likelihood ratio in likelihood inference is approximated by

$$T = 2 \log \frac{p(x | \hat{\theta})}{p(x | (\hat{\theta}_1, c_2))} \qquad (6.14)$$

when n is large. This ratio of the two observed optimal predictors can be applied to constructing a critical region. Recall that the critical region of the likelihood ratio test is constructed by applying the χ^2-approximation. Similarly, the test statistic T can be used to construct an approximated critical region.

Another likelihood inferential procedure is to evaluate a model. The model selection problem is often the selection of a prior density in the empirical Bayesian analysis. The observed optimal predictor $p_e(x|x)$ itself is not suitable for this purpose, since it uses the observation twice. Spiegelhalter et al. [18, 19] introduced the deviance information criterion (DIC), which is designed for evaluating a Bayesian model in terms of an observed predictor $p(x|x)$. It appears that they intended to discuss a general case of criteria as an extension of the AIC. Yanagimoto and Ohnishi [21] provided two specified forms of the DIC in terms of the observed e-optimal predictor. One of the two is expressed as

$$\text{uDIC} = -2 \log p(x|\hat{\theta}) + p_D + q_D, \tag{6.15}$$

where p_D and q_D denote the posterior means of $2D(p_e(y|x), p(y|\theta))$ and $2D(p(y|\theta), p_e(y|x))$, respectively. In other words, they are the posterior means of twice the dual divergence losses. The criterion in (6.15) satisfies the unbiasedness property

$$E\{2 \log p(y|\hat{\theta}) + \text{uDIC}; \; p(y|\theta)p(x|\theta)\pi(\theta)\} = 0.$$

We must emphasize that no asymptotic approximation is applied to obtain this unbiasedness property. Recall that most existing criteria were designed for approximating unbiased estimators of twice the logarithmic transformation of a plug-in predictor in terms of a suitably chosen estimator. Unfortunately, the asymptotic approximation is unavoidable in the frequentist framework, except for very simple models.

We find that the suggested procedures under Jeffreys' prior density provide asymptotically equivalent alternatives to likelihood inferential procedures. Note that this fact does not mean the superiority of one of them over the other. Thus, we hope to examine which of the two procedures performs better. Although further detailed studies will be necessary to provide recommendations, we explain reasons why Bayesian procedures are promising in the following section.

6.5 Optimality Property and Possible Improvements

We present two reasons why the procedures suggested in the previous section are expected to be superior to likelihood inferential procedures. A fact to be emphasized is that both of them may perform unfavorably. At the end of this section, we sketch our preliminary efforts to claiming that Bayesian estimators perform better in practice.

6.5.1 Optimality Property

The MLE is an efficient estimator of the mean parameter μ of the exponential family for a finite sample size n. It is also an asymptotically efficient estimator of a general parameter under weak regularity conditions. However, these optimality properties are relatively unattractive. In fact, the suggested estimator in (6.12) and its corresponding estimators are also asymptotically efficient, as shown in Corollary 6.1. The finite sample optimality is only valid, when the parameter of interest is the mean parameter and the loss is the mean squared error of μ. In addition, it does not directly relate to an optimal predictor.

In contrast, the posterior mean of θ shares asymptotic optimality properties with the MLE. In addition, it satisfies different optimality properties; an important property is that it minimizes the posterior mean of the e-divergence loss $D(p(y|x), p(y|\theta))$. This optimality property is independent of the choice of the parameter of interest. The observed predictor yields the likelihood ratio and defines the model evaluation criterion uDIC. It also minimizes the posterior mean of the squared loss $(\check{\theta} - \theta)^2$. We conjecture from these two different optimality properties that the posterior mean performs better than the MLE, though this conjecture should be confirmed by actual comparison studies.

In this concern, an important problem pertains to the fact that both the estimators can perform poorly, as stated in introduction. Two widely known pitfalls of the MLE were pointed out by Neyman and Scott [15] and James and Stein [11]. The conditional likelihood method and the marginal likelihood method were developed to overcome these problems. Jeffreys' prior density is also subject to criticism, and the posterior mean under Jeffreys' prior density should be used with caution. Critiques of Jeffreys' prior can be found in monographs such as Lindsey ([14], Chap. 6) and Robert ([16], Chap. 9).

Consequently, it is important to examine ways of improving an estimator when it is speculated as unfavorable. In the following subsections, we explore them in the case of the MLE and the posterior mean under Jeffreys' prior and observe notable differences between them.

6.5.2   Improvements on the MLE

Various methods for modifying the MLE have been proposed, including the conditional and marginal likelihood methods. However, they have been developed in order to eliminate the nuisance parameter; see Lindsey ([14], Chap. 6), for example. More seriously, such modifications depend on specific forms of the sampling density. Indeed, the conditional method requires the existence of a statistic t such that the conditional likelihood given t is independent of the remaining part of the parameter θ. Thus, its application is severely restricted. Another reservation arises because an estimator induced from such a modification does not attain the maximum of the likelihood, and so it invalidates the maximum likelihood principle. See Lehmann ([13], p.17) and Goodfellow et al. ([10], p.128) for this principle. There exist some attempts to extend the conditional likelihood method (e.g., [6]), but such attempts still require strong assumptions on the structure of the sampling density.

6.5.3 Improvements on the Posterior Mean Under Jeffreys' Prior

In contrast to the likelihood method, Bayesian methods are simple and flexible, requiring only the assumption of a suitable prior density. Although the assumption of Jeffreys' prior density is useful for evaluating the MLE, we can choose another suitable prior density from a wide variety of possible priors.

Other non-informative prior densities, such as the reference prior, may be assumed if a non-informative prior density is desired. It is recommended to assume an informative prior density, if possible. When the sampling density is in the exponential family, a conjugate prior density is a promising candidate. An informative prior density is proper, and its optimality properties hold in a general way. For example, the posterior mean of θ minimizes the posterior mean of the squared error $(\check{\theta} - \theta)^2$ as well as the e-divergence loss.

Apart from the conjugate prior density, it is possible to apply the empirical Bayes model. However, such sophisticated models are not necessary for our comparison with the likelihood method because a variety of Bayesian procedures are induced from conjugate prior densities based on Jeffreys' prior density, even when it is subject to criticism.

We close this section by presenting our preliminary attempts to claim the better performance of Bayesian estimators in an explicit class of distributions in the exponential family. Consider, for example, the family of von-Mises distributions with the location and the dispersion parameters, which is familiar in analyzing circular data (e.g., [8]). The MLE and the maximum marginal likelihood estimator are two standard estimators in the frequentist context, and the latter estimator performs better than the former. In this model, neither the posterior mean of the canonical parameter under Jeffreys' prior nor the MLE performs favorably. Jeffreys' prior can thus be replaced by the reference prior. Preliminary comparison studies suggest that the posterior mean of the canonical parameter under Jeffreys' prior takes smaller estimated risks than the MLE for a wide range of parameters [17]. Similar results are observed between the posterior means under Jeffreys' prior and the MLE as well as between that under the reference prior and the conditional MLE in the inverse Gaussian distribution.

6.6 Discussion and Recommendations

Our claims here consist of two parts: (i) that the inferential procedures induced from the e-optimal predictor $p_e(y|x)$ are promising, and (ii) that the posterior mean of the canonical parameter takes a value close to the MLE. The former assertion is supported by the fact that they satisfy favorable properties. The plug-in predictive density $p(y|\hat{\theta})$ is in the family of sampling densities $\mathcal{F}$, which satisfies an optimality property. The Bayesian statistics corresponding to the likelihood ratio and the AIC can be defined, as in Sect. 6.4.

The close relationship between the posterior mean of the canonical parameter under Jeffreys' prior and the MLE is useful for better understanding these two estimators. Indeed, we can expect that the MLE performs favorably when Jeffreys' prior for a parameter is acceptable and vice versa. Thus, we can examine the validity of the MLE and Jeffreys' prior in parallel ways. On the other hand, it is worthwhile investigating which of these two estimators actually performs better in a specified family.

The above claims lead us to the following two recommendations. First, consider the case in which Jeffreys' prior density is unacceptable. The behavior of the MLE allows us to examine this judgement. The assumption of another non-informative prior density or an informative prior density is recommended. This recommendation may be trivial from Bayesian standpoints, but the suggestion regarding the behavior of the MLE is practically useful. Another recommendation is to explore the procedures induced from the e-optimal predictor. Note that the simple form of μ^* in (6.3) is not sufficient for constructing practical procedures.

Acknowledgements The authors express their thanks to a reviewer and the editors for their comments on points to be clarified.

Appendix A. Proof of Theorem 6.1

Before presenting the proof, we clarify the notation that is more rigorous than that in the text. Write a density in the exponential family as

$$p(x|\theta) = \prod \exp\{\theta \cdot t_i - M(\theta)\}a(x_i),$$

where $\sum t_i \in \mathcal{X} \subset \mathbb{R}^p$ is the sufficient statistic, and $\theta \in \Theta \subset \mathbb{R}^p$ with $\theta = (\theta_1, \ldots, \theta_p)$ is the canonical parameter.

For a given θ_0, the corresponding mean parameter is written as $\mu_0 = \nabla M(\theta_0)$. The Kullback–Leibler divergence is expressed as

$$D(\theta_0, \theta) = M(\theta) + N(\mu_0) - \mu_0 \cdot \theta,$$

and the Fisher information matrix is written as

$$I(\theta) = \{M_{ij}(\theta)\}_{1 \le i, j \le p},$$

where $M_{ij}(\theta) = \partial^2 M(\theta)/\partial\theta_i\partial\theta_j$. For notational convenience, the partial derivative of a function of a vector variable with respect to its components is denoted by the corresponding suffixes.

We begin the proof with presenting an expression of the posterior mean as

$$E\{\theta\ ;\ \pi(\theta;\theta_0,n)\} = \frac{\int_\Theta \theta b(\theta)\exp\{-nD(\theta_0,\theta)\}d\theta}{\int_\Theta b(\theta)\exp\{-nD(\theta_0,\theta)\}d\theta}.$$

Thus, we may evaluate

$$\int_\Theta g(\theta)\exp\{-nD(\theta_0,\theta)\}d\theta \tag{6.16}$$

for cases in which $g(\theta)$ is $\theta_i b(\theta)$ and $b(\theta)$.

When $\theta \approx \theta_0$, the following formal approximation is possible:

$$\begin{aligned}
D(\theta_0,\theta) &\approx \frac{1}{2}(\theta-\theta_0)^T I(\theta_0)(\theta-\theta_0) \\
&+ \frac{1}{3!}\sum_{j_1,j_2,j_3} M_{j_1 j_2 j_3}(\theta_0)(\theta_{j_1}-\theta_{0j_1})(\theta_{j_2}-\theta_{0j_2})(\theta_{j_3}-\theta_{0j_3}) \\
&+ \frac{1}{4!}\sum_{j_1,j_2,j_3,j_4} M_{j_1 j_2 j_3 j_4}(\theta_0)(\theta_{j_1}-\theta_{0j_1})(\theta_{j_2}-\theta_{0j_2})(\theta_{j_3}-\theta_{0j_3})(\theta_{j_4}-\theta_{0j_4}),
\end{aligned}$$

where θ_{0i} denotes the i-th component of θ_0.

Since $I(\theta_0)$ is assumed to be positive definite, the a-th power can be defined for $a = 1/2$ and $-1/2$. Both matrices are positive definite, and one is the inverse matrix of the other. We consider here the following parameter transformation of θ to z as

$$z = \sqrt{n}I^{1/2}(\theta_0)(\theta-\theta_0).$$

The Jacobian of this transformation is

$$\frac{1}{n^{p/2}\sqrt{\det I(\theta_0)}}.$$

Then the asymptotic expansion of the Kullback–Leibler divergence up to the order $O(1/n)$ is given by

$$nD(\theta_0,\theta) \approx \frac{|z|^2}{2} + \frac{1}{\sqrt{n}}\sum_{j_1,j_2,j_3} a^{(1)}_{j_1 j_2 j_3} z_{j_1} z_{j_2} z_{j_3} + \frac{1}{n}\sum_{j_1,j_2,j_3,j_4} a^{(2)}_{j_1 j_2 j_3 j_4} z_{j_1} z_{j_2} z_{j_3} z_{j_4},$$

where $a^{(1)}_{j_1 j_2 j_3}$ and $a^{(2)}_{j_1 j_2 j_3 j_4}$ are defined as

$$a^{(1)}_{j_1 j_2 j_3} := \frac{1}{3!}\sum_{j_4,j_5,j_6} M_{j_4 j_5 j_6}(\theta_0) I^{-1/2}_{j_4 j_1}(\theta_0) I^{-1/2}_{j_5 j_2}(\theta_0) I^{-1/2}_{j_6 j_3}(\theta_0); \tag{6.17}$$

$$a^{(2)}_{j_1 j_2 j_3 j_4} := \frac{1}{4!} \sum_{j_5, j_6, j_7, j_8} M_{j_5 j_6 j_7 j_8}(\theta_0) I^{-1/2}_{j_5 j_1}(\theta_0) I^{-1/2}_{j_6 j_2}(\theta_0) I^{-1/2}_{j_7 j_3}(\theta_0) I^{-1/2}_{j_8 j_4}(\theta_0). \quad (6.18)$$

Note that $a^{(1)}_{j_1 j_2 j_3}$ and $a^{(2)}_{j_1 j_2 j_3 j_4}$ remain unchanged under the permutation of the suffixes, since $M(\theta)$ is assumed to be of C^4 class. To evaluate the integral in (6.16), we evaluate the asymptotic expansion of $\exp\{-nD(\theta_0, \theta)\}$. Writing the density of the standard p-dimensional normal as $\phi(z)$, we can give the asymptotic expansion up to the order $O(1/n)$ as

$$\exp\{-nD(\theta_0, \theta)\}$$

$$= (2\pi)^{p/2}\phi(z) \exp\left(-\frac{1}{\sqrt{n}} \sum_{j_1, j_2, j_3} a^{(1)}_{j_1 j_2 j_3} z_{j_1} z_{j_2} z_{j_3} - \frac{1}{n} \sum_{j_1, j_2, j_3, j_4} a^{(2)}_{j_1 j_2 j_3 j_4} z_{j_1} z_{j_2} z_{j_3} z_{j_4} \right)$$

$$\approx (2\pi)^{p/2}\phi(z)\left(1 - \frac{1}{\sqrt{n}} \sum_{j_1, j_2, j_3} a^{(1)}_{j_1 j_2 j_3} z_{j_1} z_{j_2} z_{j_3} - \frac{1}{n} \sum_{j_1, j_2, j_3, j_4} a^{(2)}_{j_1 j_2 j_3 j_4} z_{j_1} z_{j_2} z_{j_3} z_{j_4} \right.$$

$$\left. + \frac{1}{2n} \sum_{j_1, j_2, j_3, j_4, j_5, j_6} a^{(1)}_{j_1 j_2 j_3} a^{(1)}_{j_4 j_5 j_6} z_{j_1} z_{j_2} z_{j_3} z_{j_4} z_{j_5} z_{j_6} \right).$$

In a sequel, we regard the domain of θ as $\mathbb{R}^p$.

Next, we calculate the asymptotic expansion of $g(\theta)$ up to the order $O(1/n)$ by

$$g(\theta) \approx g(\theta_0) + \frac{1}{\sqrt{n}} \sum_{j_1, j_2} g_{j_2}(\theta_0) I^{-1/2}_{j_2 j_1}(\theta_0) z_{j_1}$$

$$+ \frac{1}{2n} \sum_{j_1, j_2, j_3, j_4} g_{j_3, j_4}(\theta_0) I^{-1/2}_{j_3 j_1}(\theta_0) I^{-1/2}_{j_4 j_2}(\theta_0) z_{j_1} z_{j_2}$$

$$= g(\theta_0)\left(1 + \frac{1}{\sqrt{n}} \sum_{j_1} c^{(1)}_{j_1} z_{j_1} + \frac{1}{n} \sum_{j_1, j_2} c^{(2)}_{j_1 j_2} z_{j_1} z_{j_2} \right),$$

where $c^{(1)}_{j_1}$ and $c^{(2)}_{j_1 j_2}$ denote, respectively,

$$c^{(1)}_{j_1} := \sum_{j_2} \frac{g_{j_2}(\theta_0)}{g(\theta_0)} I^{-1/2}_{j_2 j_1}(\theta_0) \quad (6.19)$$

and

$$c^{(2)}_{j_1 j_2} := \frac{1}{2} \sum_{j_3, j_4} \frac{g_{j_3 j_4}(\theta_0)}{g(\theta_0)} I^{-1/2}_{j_3 j_1}(\theta_0) I^{-1/2}_{j_4 j_2}(\theta_0). \quad (6.20)$$

Note that these coefficients also remain unchanged under the permutation of the suffixes, as are $a^{(1)}_{j_1 j_2 j_3}$ and $a^{(2)}_{j_1 j_2 j_3 j_4}$ in (6.17) and (6.18).

Combining these asymptotic expansions, we obtain that of the integrant in (6.16) as follows:

$$g(\theta)\exp\{-nD(\theta_0,\theta)\}$$

$$\approx (2\pi)^{p/2}g(\theta_0)\phi(z)\left(1 + \frac{1}{\sqrt{n}}\sum_{j_1}c_{j_1}^{(1)}z_{j_1} + \frac{1}{n}\sum_{j_1,j_2}c_{j_1 j_2}^{(2)}z_{j_1}z_{j_2}\right)$$

$$\times\left(1 - \frac{1}{\sqrt{n}}\sum_{j_1,j_2,j_3}a_{j_1 j_2 j_3}^{(1)}z_{j_1}z_{j_2}z_{j_3} - \frac{1}{n}\sum_{j_1,j_2,j_3,j_4}a_{j_1 j_2 j_3 j_4}^{(2)}z_{j_1}z_{j_2}z_{j_3}z_{j_4}\right.$$

$$\left. + \frac{1}{2n}\sum_{j_1,j_2,j_3,j_4,j_5,j_6}a_{j_1 j_2 j_3}^{(1)}a_{j_4 j_5 j_6}^{(1)}z_{j_1}z_{j_2}z_{j_3}z_{j_4}z_{j_5}z_{j_6}\right).$$

Since this approximated integrant contains $\phi(z)$, we may discard the odd order terms of the polynomial of z to give

$$(2\pi)^{p/2}g(\theta_0)\phi(z)$$

$$\times\left\{1 - \frac{1}{n}\sum_{j_1,j_2,j_3,j_4}a_{j_1 j_2 j_3 j_4}^{(2)}z_{j_1}z_{j_2}z_{j_3}z_{j_4}\right.$$

$$+ \frac{1}{2n}\sum_{j_1,j_2,j_3,j_4,j_5,j_6}a_{j_1 j_2 j_3}^{(1)}a_{j_4 j_5 j_6}^{(1)}z_{j_1}z_{j_2}z_{j_3}z_{j_4}z_{j_5}z_{j_6}$$

$$\left. - \frac{1}{n}\sum_{j_1,j_2,j_3,j_4}c_{j_1}^{(1)}a_{j_2 j_3 j_4}^{(1)}z_{j_1}z_{j_2}z_{j_3}z_{j_4} + \frac{1}{n}\sum_{j_1,j_2}c_{j_1 j_2}^{(2)}z_{j_1}z_{j_2}\right\}.$$

Let $Z = (Z_1, \ldots, Z_p)^T$ be a random variable having the density $\phi(z)$. Then the second moment is written as $E\{Z_i Z_j\} = \delta_{ij}$. To evaluate the fourth moment, set $\gamma_{ijkl} := E\{Z_i Z_j Z_k Z_l\}$. It follows that $\gamma_{iiii} = 3$ for every i, and that $\gamma_{iikk} = 1$ for (i, j, k, l) such that two pairs take different integers. The sixth moment remains in the asymptotic expansion of the integral of (6.16), but disappears in the asymptotic expansion of the posterior mean.

The asymptotic expansion of the integral of (6.16) up to the order $O(1/n)$ is expressed as

$$\int_{\Theta}g(\theta)\exp\{-nD(\theta_0,\theta)\}d\theta$$

$$\approx \left(\frac{2\pi}{n}\right)^{p/2}\frac{g(\theta_0)}{\sqrt{\det I(\theta_0)}}$$

$$\times\left\{1 - \frac{1}{n}\sum_{j_1,j_2,j_3,j_4}a_{j_1 j_2 j_3 j_4}^{(2)}\gamma_{j_1 j_2 j_3 j_4} + \frac{1}{2n}\sum_{j_1,j_2,j_3,j_4,j_5,j_6}a_{j_1 j_2 j_3}^{(1)}a_{j_4 j_5 j_6}^{(1)}\gamma_{j_1 j_2 j_3 j_4 j_5 j_6}\right.$$

$$-\frac{1}{n}\sum_{j_1,j_2,j_3,j_4} c^{(1)}_{j_1} a^{(1)}_{j_2 j_3 j_4} \gamma_{j_1 j_2 j_3 j_4} + \frac{1}{n}\sum_{j_1,j_2} c^{(2)}_{j_1 j_2} \delta_{j_1 j_2}\Bigg\}.$$

Set the sum of the second and third terms as A, that is,

$$A = -\sum_{j_1,j_2,j_3,j_4} a^{(2)}_{j_1 j_2 j_3 j_4}\gamma_{j_1 j_2 j_3 j_4} + \frac{1}{2}\sum_{j_1,j_2,j_3,j_4,j_5,j_6} a^{(1)}_{j_1 j_2 j_3} a^{(1)}_{j_4 j_5 j_6}\gamma_{j_1 j_2 j_3 j_4 j_5 j_6}.$$

Note that A is independent of $b(\theta)$. Next, we simplify the fourth term, by applying the properties of $a^{(1)}_{j_1 j_2 j_3}$, as follows:

$$\sum_{j_1,j_2,j_3,j_4} c^{(1)}_{j_1} a^{(1)}_{j_2 j_3 j_4}\gamma_{j_1 j_2 j_3 j_4}$$

$$= 3\sum_{j_1} c^{(1)}_{j_1} a^{(1)}_{j_1 j_1 j_1} + \sum_{j_1\neq j_2} c^{(1)}_{j_1} a^{(1)}_{j_1 j_2 j_2} + \sum_{j_1\neq j_2} c^{(1)}_{j_1} a^{(1)}_{j_2 j_1 j_2} + \sum_{j_1\neq j_2} c^{(1)}_{j_1} a^{(1)}_{j_2 j_2 j_1}$$

$$= 3\sum_{j_1} c^{(1)}_{j_1} a^{(1)}_{j_1 j_1 j_1} + 3\sum_{j_1\neq j_2} c^{(1)}_{j_1} a^{(1)}_{j_1 j_2 j_2}$$

$$= 3\sum_{j_1,j_2} c^{(1)}_{j_1} a^{(1)}_{j_1 j_2 j_2}.$$

The fifth term is rewritten as

$$\sum_{j_1,j_2} c^{(2)}_{j_1 j_2}\delta_{j_1 j_2} = \sum_{j_1} c^{(2)}_{j_1 j_1}.$$

Therefore, the asymptotic expansion of the integral is of the form

$$\int_\Theta g(\theta)\exp\{-n\mathrm{D}(\theta_0,\theta)\}d\theta$$

$$\approx \left(\frac{2\pi}{n}\right)^{p/2}\frac{g(\theta_0)}{\sqrt{\det I(\theta_0)}}\left\{1 + \frac{1}{n}\left(A - 3\sum_{j_1,j_2} c^{(1)}_{j_1} a^{(1)}_{j_1 j_2 j_2} + \sum_{j_1} c^{(2)}_{j_1 j_1}\right)\right\}.$$

Thus, both the numerator and the denominator of the posterior mean are expressed in similar forms as

$$\int_\Theta \theta_i b(\theta)\exp\{-n\mathrm{D}(\theta_0,\theta)\}d\theta \approx \left(\frac{2\pi}{n}\right)^{p/2}\frac{\theta_{0i} g(\theta_0)}{\sqrt{\det I(\theta_0)}}\left(1 + \frac{d_{Ni}}{n}\right),$$

$$\int_\Theta b(\theta)\exp\{-n\mathrm{D}(\theta_0,\theta)\}d\theta \approx \left(\frac{2\pi}{n}\right)^{p/2}\frac{g(\theta_0)}{\sqrt{\det I(\theta_0)}}\left(1 + \frac{d_D}{n}\right).$$

Consequently, we obtain the asymptotic expansion of the posterior mean as

$$\mathrm{E}\{\theta_i \; ; \; \pi(\theta; \theta_0, n)\} \approx \theta_{0i}\left(1 + \frac{d_{Ni} - d_D}{n}\right).$$

Next, we prove the necessity and suppose that $d_{Ni} = d_D$ for every i. Since the coefficients $a^{(1)}_{j_1 j_2 j_3}$ and $a^{(2)}_{j_1 j_2 j_3 j_4}$ are independent of $b(\theta)$, the difference $d_{Ni} = d_D$ is independent of A for every i. Thus, the difference depends on $c^{(1)}_{j_1}$ and $c^{(2)}_{j_1 j_1}$. To evaluate the difference, we decompose it into two terms $F_1 + F_2$ such that F_1 is a function of $c^{(1)}_{j_1}$ and F_2 is a function of $c^{(2)}_{j_1 j_1}$.

Since $M(\theta)$ is assumed to be of C^4 class, $I(\theta)$ is of C^2 class. Using the equality

$$\frac{1}{\theta_i b(\theta)}\frac{\partial\{\theta_i b(\theta)\}}{\partial\theta_{j_2}} - \frac{b_{j_2}(\theta)}{b(\theta)} = \frac{\delta_{j_2 i}}{\theta_i},$$

we find that the coefficient of $c^{(1)}_{j_1}$ in the difference $d_{Ni} = d_D$ is written as

$$\sum_{j_3}\frac{\delta_{j_3 i}}{\theta_{0i}}I^{-1/2}_{j_3 j_1}(\theta_0) = \frac{I^{-1/2}_{ij_1}(\theta_0)}{\theta_{0i}}.$$

This implies that F_1 can be expressed as follows:

$$F_1 = -3\sum_{j_1 j_2}\frac{I^{-1/2}_{ij_1}(\theta_0)}{\theta_{0i}}a^{(1)}_{j_1 j_2 j_2}$$

$$= -3\sum_{j_1 j_2}\frac{I^{-1/2}_{ij_1}(\theta_0)}{\theta_{0i}}\frac{1}{3!}\sum_{j_4, j_5, j_6}M_{j_4 j_5 j_6}(\theta_0)I^{-1/2}_{j_4 j_1}(\theta_0)I^{-1/2}_{j_5 j_2}(\theta_0)I^{-1/2}_{j_6 j_2}(\theta_0)$$

$$= -\frac{1}{2\theta_{0i}}\sum_{j_1, j_2, j_4, j_5, j_6}I^{-1/2}_{ij_1}(\theta_0)I^{-1/2}_{j_4 j_1}(\theta_0)I^{-1/2}_{j_5 j_2}(\theta_0)I^{-1/2}_{j_6 j_2}(\theta_0)M_{j_4 j_5 j_6}(\theta_0).$$

Since $I^{-1/2}(\theta_0)$ is symmetric, it follows that

$$\sum_{j_1}I^{-1/2}_{ij_1}(\theta_0)I^{-1/2}_{j_4 j_1}(\theta_0) = \sum_{j_1}I^{-1/2}_{ij_1}(\theta_0)I^{-1/2}_{j_1 j_4}(\theta_0) = I^{-1}_{ij_4}(\theta_0)$$

and also that

$$\sum_{j_2}I^{-1/2}_{j_5 j_2}(\theta_0)I^{-1/2}_{j_6 j_2}(\theta_0) = I^{-1}_{j_5 j_6}(\theta_0).$$

Consequently, the former term is given by

$$F_1 = -\frac{1}{2\theta_{0i}} \sum_{j_4, j_5, j_6} I_{ij_4}^{-1}(\theta_0) I_{j_5 j_6}^{-1}(\theta_0) M_{j_4 j_5 j_6}(\theta_0).$$

Next, we evaluate the latter term F_2. It holds that

$$\frac{1}{\theta_i b(\theta)} \frac{\partial^2 \{\theta_i b(\theta)\}}{\partial \theta_{j_3} \partial_{j_4}} - \frac{b_{j_3 j_4}(\theta)}{b(\theta)} = \frac{\delta_{j_4 i} b_{j_3}(\theta) + \delta_{j_3 i} b_{j_4}(\theta)}{\theta_i b(\theta)}.$$

Hence, the coefficient of $c_{j_1 j_1}^{(2)}$ in the difference $d_{Ni} = d_D$ is written as

$$\frac{1}{2} \sum_{j_3, j_4} \frac{\delta_{j_4 i} b_{j_3}(\theta_0) + \delta_{j_3 i} b_{j_4}(\theta_0)}{\theta_{0i} b(\theta)} I_{j_3 j_1}^{-1/2}(\theta_0) I_{j_4 j_1}^{-1/2}(\theta_0)$$

$$= \frac{I_{ij_1}^{-1/2}(\theta_0)}{2\theta_{0i}} \sum_{j_3} \frac{b_{j_3}(\theta_0)}{b(\theta)} I_{j_3 j_1}^{-1/2}(\theta_0) + \frac{I_{ij_1}^{-1/2}(\theta_0)}{2\theta_{0i}} \sum_{j_4} \frac{b_{j_4}(\theta_0)}{b(\theta)} I_{j_4 j_1}^{-1/2}(\theta_0)$$

$$= \frac{I_{ij_1}^{-1/2}(\theta_0)}{\theta_{0i}} \sum_{j_3} \frac{b_{j_3}(\theta_0)}{b(\theta)} I_{j_3 j_1}^{-1/2}(\theta_0).$$

Thus, the latter term F_2 is given by

$$F_2 = \frac{1}{\theta_{0i}} \sum_{j_1, j_3} I_{ij_1}^{-1/2}(\theta_0) \frac{b_{j_3}(\theta_0)}{b(\theta)} I_{j_3 j_1}^{-1/2}(\theta_0) = \frac{1}{\theta_{0i}} \sum_{j_3} I_{ij_3}^{-1}(\theta_0) \frac{b_{j_3}(\theta_0)}{b(\theta)}.$$

Combining these results, we obtain that

$$d_{Ni} - d_D = -\frac{1}{2\theta_{0i}} \sum_{j_4, j_5, j_6} I_{ij_4}^{-1}(\theta_0) I_{j_5 j_6}^{-1}(\theta_0) M_{j_4 j_5 j_6}(\theta_0) + \frac{1}{\theta_{0i}} \sum_{j_3} I_{ij_3}^{-1}(\theta_0) \frac{b_{j_3}(\theta_0)}{b(\theta)}.$$

$$(6.21)$$

Using the equality

$$M_{j_3 j_5 j_6}(\theta_0) = \frac{\partial I_{j_5 j_6}(\theta_0)}{\partial \theta_{0 j_3}}$$

and replacing the index j_4 in the summation of the difference (6.21) by j_3, we can rewrite the difference as follows:

$$d_{Ni} - d_D = \frac{1}{\theta_{0i}} \sum_{j_3} I_{ij_3}^{-1}(\theta_0) \left\{ \frac{b_{j_3}(\theta_0)}{b(\theta)} - \frac{1}{2} \sum_{j_5, j_6} I_{j_5 j_6}^{-1}(\theta_0) \frac{\partial I_{j_5 j_6}(\theta_0)}{\partial \theta_{0 j_3}} \right\}.$$

Applying the differentiation formula for the determinant of a differentiable and invertible matrix $A(t)$, $d\{A(t)\}/dt = \det A(t) \, \text{tr}\{A^{-1} d\{A(t)\}/dt\}$, we can express the right-hand side in terms of the derivative of the determinant of the matrix $I(\theta_0)$:

$$\sum_{j_5, j_6} I_{j_5 j_6}^{-1}(\theta_0) \frac{\partial I_{j_5 j_6}(\theta_0)}{\partial \theta_{0 j_3}} = \frac{\partial}{\partial \theta_{0 j_3}} \log \det I(\theta_0).$$

This implies that

$$d_{Ni} - d_D = \frac{1}{\theta_{0i}} \sum_{j_3} I_{i j_3}^{-1}(\theta_0) \frac{\partial}{\partial \theta_{0 j_3}} \left\{ \log b(\theta) - \frac{1}{2} \log \det I(\theta_0) \right\}.$$

The condition for this equality to hold for every i is expressed as

$$\nabla \log \frac{b(\theta)}{\sqrt{\det I(\theta)}} = 0.$$

This completes the proof.

Appendix B. Proof of Corollary 6.1

To apply Theorem 6.1, set $s = \nabla N(t)$, and define the function $f_n(s)$ as

$$f_n(s) = \frac{\int \theta \exp\{-n D(s, \theta)\} \pi_J(\theta) d\theta}{\int \exp\{-n D(s, \theta)\} \pi_J(\theta) d\theta}.$$

Theorem 6.1 yields that, for an arbitrary fixed s,

$$f_n(s) = s + a_n(s)/n^2, \tag{6.22}$$

where the coefficient $a_n(s)$ is continuous and is of the order $O(1)$.

Write the MLE of θ for a sample of size n, x_n, as $\hat{\theta}_{ML}(x_n)$. Since the sample density is assumed to be in the exponential family, $\hat{\theta}_{ML}(x_n)$ can be expressed as $\nabla N(\bar{t})$, which is written as s_n. Then, the posterior mean $\hat{\theta}(x_n)$ is expressed as $\hat{\theta}(x_n) = f_n(s_n)$. The assumption that the sampling density is in the regular exponential family shows that a true parameter θ_T is in the interior of Θ; the law of large numbers implies that for an arbitrary small positive value ϵ, the probability of the subspace of samples $\mathcal{X}_\epsilon(n) = \{x_n | |s_n - \theta_T| \le \epsilon\}$ is greater than $1 - \epsilon$. From the continuity of the coefficient $a_n(s_n)$ in (6.22), it follows that $a_n(s_n)$ is bounded for $x_n \in \mathcal{X}_\epsilon(n)$. Thus, it holds that

$$f_n(s_n) = s_n + c_n/n^2,$$

where c_n is a finite value. Combining these results, we find that

$$\hat{\theta}(x_n) = \hat{\theta}_{ML}(x_n) + O_P(1/n^2).$$

References

1. Aitchison J (1975) Goodness of prediction fit. Biometrika 62:547–554
2. Amari S, Nagaoka H (2007) Methods of information geometry. Am Math Soc, Rhode Island
3. Berger JO, Bernardo JM (1992) Ordered group reference priors with application to the multinomial problem. Biometrika 79:25–37
4. Bernardo JM (1979) Reference posterior distributions for Bayesian inference. J Roy Statist Soc B 41:113–147
5. Corcuera JM, Giummole F (1999) A generalized Bayes rule for prediction. Scand J Statist 26:265–279
6. Cox DR, Reid N (1987) Parameter orthogonality and approximate conditional inference (with discussion). J Roy Statist Soc B 49:1–39
7. Diaconis P, Ylvisaker D (1979) Conjugate priors for exponential families. Ann Statist 7:269–281
8. Fisher NL (1995) Statistical Analysis of circular data. Cambridge University Press, Cambridge
9. Ghosh M, Liu R (2011) Moment matching priors. Sankhya A 73:185–201
10. Goodfellow I, Bengio Y, Courville A (2016) Deep learning. MIT Press, Cambridge
11. James W, Stein C (1961) Estimation with quadratic loss. Proc Fourth Berkeley Symp Math Statist Prob 1:361–380
12. Jeffreys H (1961) Theory of probability, 3rd edn. Oxford Univ Press, Oxford
13. Lehmann EL (1959) Testing statistical hypotheses. Wiley, New York
14. Lindsey JK (1996) Parametric statistical inference. Clarendon Press, Oxford
15. Neyman J, Scott EL (1948) Consistent estimates based on partially consistent observations. Econometrica 16:1–32
16. Robert CP (2001) The Bayesian choice, 2nd edn. Springer, New York
17. Sakumura T and Yanagimoto T (2019) Posterior mean of the canonical parameter in the von-Mises distribution (in Japanese). Read at Japan Joint Statist Meet Abstract 69
18. Spiegelhalter DJ, Best NG, Carlin BP, van der Lind A (2002) Bayesian measures of model complexity and fit (with discussions). J R Statist Soc B 64:583–639
19. Spiegelhalter DJ, Best NG, Carlin BP, van der Linde A (2014) The deviance information criterion: 12 years on. J R Statist Soc B 76:485–493
20. Tierney L, Kass RE, Kadane JB (1989) Fully exponential Laplace approximations to expectations and variances of nonpositive functions. J Am Statist Assoc 84:710–716
21. Yanagimoto T, Ohnishi T (2009) Bayesian prediction of a density function in terms of e-mixture. J Statist Plann Inf 139:3064–3075
22. Yanagimoto T, Ohnishi T (2011) Saddlepoint condition on a predictor to reconfirm the need for the assumption of a prior distribution. J Statist Plann Inf 41:1990–2000

Printed in the United States
By Bookmasters